> VOLUME SIXTY NINE

ADVANCES IN
CATALYSIS

EDITOR IN CHIEF

M. DIÉGUEZ
Universitat Rovira i Virgili, Tarragona, Spain

ADVISORY BOARD

A. CORMA CANÓS
Valencia, Spain

G. ERTL
Berlin/Dahlem, Germany

B.C. GATES
Davis, California, USA

G. HUTCHINGS
Cardiff, UK

E. IGLESIA
Berkeley, California, USA

P.W.N.M. VAN LEEUWEN
Toulouse, France

J. ROSTRUP-NIELSEN
Lyngby, Denmark

R.A. VAN SANTEN
Eindhoven, The Netherlands

F. SCHÜTH
Mülheim, Germany

J.M. THOMAS
London/Cambridge, England

VOLUME SIXTY NINE

ADVANCES IN
CATALYSIS

Edited by

MONTSERRAT DIÉGUEZ
Departament de Química Física i Inorgànica,
Universitat Rovira i Virgili, Tarragona, Spain

OSCAR PÀMIES
Departament de Química Física i Inorgànica,
Universitat Rovira i Virgili, Tarragona, Spain

Academic Press is an imprint of Elsevier
50 Hampshire Street, 5th Floor, Cambridge, MA 02139, United States
525 B Street, Suite 1650, San Diego, CA 92101, United States
The Boulevard, Langford Lane, Kidlington, Oxford OX5 1GB, United Kingdom
125 London Wall, London, EC2Y 5AS, United Kingdom

First edition 2021

Copyright © 2021 Elsevier Inc. All rights reserved.

No part of this publication may be reproduced or transmitted in any form or by any means, electronic or mechanical, including photocopying, recording, or any information storage and retrieval system, without permission in writing from the publisher. Details on how to seek permission, further information about the Publisher's permissions policies and our arrangements with organizations such as the Copyright Clearance Center and the Copyright Licensing Agency, can be found at our website: www.elsevier.com/permissions.

This book and the individual contributions contained in it are protected under copyright by the Publisher (other than as may be noted herein).

Notices
Knowledge and best practice in this field are constantly changing. As new research and experience broaden our understanding, changes in research methods, professional practices, or medical treatment may become necessary.

Practitioners and researchers must always rely on their own experience and knowledge in evaluating and using any information, methods, compounds, or experiments described herein. In using such information or methods they should be mindful of their own safety and the safety of others, including parties for whom they have a professional responsibility.

To the fullest extent of the law, neither the Publisher nor the authors, contributors, or editors, assume any liability for any injury and/or damage to persons or property as a matter of products liability, negligence or otherwise, or from any use or operation of any methods, products, instructions, or ideas contained in the material herein.

ISBN: 978-0-12-824571-2
ISSN: 0360-0564

For information on all Academic Press publications
visit our website at https://www.elsevier.com/books-and-journals

Publisher: Zoe Kruze
Acquisitions Editor: Sam Mahfoudh
Developmental Editor: Tara Nadera
Production Project Manager: Denny Mansingh
Cover Designer: Victoria Pearson

Typeset by STRAIVE, India

Contents

Contributors	*vii*
Preface	*ix*

1. Metal-catalyzed biomimetic aerobic oxidation of organic substrates   1

Srimanta Manna, Wei-Jun Kong, and Jan-E. Bäckvall

1. Introduction	2
2. Metal-catalyzed biomimetic aerobic oxidations by multistep electron transfer	5
3. Aerobic oxidation by biomimetic metal complexes	41
4. Concluding remarks and future perspectives	51
Acknowledgments	52
References	52
About the authors	56

2. Zeolites catalyze selective reactions of large organic molecules   59

Marta Mon and Antonio Leyva-Pérez

1. Introduction	60
2. Metathesis reaction	62
3. Hydroaddition reactions to alkenes and alkynes	69
4. Electrocyclization reactions	85
5. Imination reactions	88
6. Selective alkylation reactions	89
7. Conclusions and outlook	96
Acknowledgments	97
References	97
About the authors	102

3. Metal-π-allyl mediated asymmetric cycloaddition reactions   103

Pol de la Cruz-Sánchez and Oscar Pàmies

1. Introduction	104
2. Cycloaddition reactions for the formation of *O*-heterocycles	105
3. Cycloaddition reactions for the formation of *N*-heterocycles	124
4. Cycloaddition reactions for the formation of carbocycles	145
5. Cycloaddition reactions for the formation of mixed-heterocycles	163

6. Conclusions	171
Acknowledgments	171
References	172
About the authors	180

4. Evolution in the metal-catalyzed asymmetric hydroformylation of 1,1′-disubstituted alkenes
181

Jèssica Margalef, Joris Langlois, Guillem Garcia, Cyril Godard, and Montserrat Diéguez

1. Introduction	182
2. Regioselectivity on the asymmetric hydroformylation of 1,1′-disubstituted alkenes	184
3. Asymmetric hydroformylation of 1,1′-disubsituted alkenes with coordinative groups	185
4. Asymmetric hydroformylation of 1,1′-disubsituted alkenes with non-coordinative groups	199
5. Rationalization of the catalyst efficiency	207
6. Conclusions and outlook	210
Acknowledgments	211
References	211
About the authors	213

Contributors

Jan-E. Bäckvall
Department of Organic Chemistry, Arrhenius Laboratory, Stockholm University, Stockholm, Sweden

Pol de la Cruz-Sánchez
Universitat Rovira i Virgili, Departament de Química Física i Inorgànica, C/Marcel·lí Domingo, Tarragona, Spain

Montserrat Diéguez
Departament de Química Física i Inorgànica, Universitat Rovira i Virgili, C/Marcel·lí Domingo, Tarragona, Spain

Guillem Garcia
Departament de Química Física i Inorgànica, Universitat Rovira i Virgili, C/Marcel·lí Domingo, Tarragona, Spain

Cyril Godard
Departament de Química Física i Inorgànica, Universitat Rovira i Virgili, C/Marcel·lí Domingo, Tarragona, Spain

Wei-Jun Kong
Department of Organic Chemistry, Arrhenius Laboratory, Stockholm University, Stockholm, Sweden

Joris Langlois
Departament de Química Física i Inorgànica, Universitat Rovira i Virgili, C/Marcel·lí Domingo, Tarragona, Spain

Antonio Leyva-Pérez
Instituto de Tecnología Química (UPV–CSIC), Universitat Politècnica de València–Consejo Superior de Investigaciones Científicas Avda. de los Naranjos s/n, Valencia, Spain

Srimanta Manna
Department of Organic Chemistry, Arrhenius Laboratory, Stockholm University, Stockholm, Sweden

Jèssica Margalef
Departament de Química Física i Inorgànica, Universitat Rovira i Virgili, C/Marcel·lí Domingo, Tarragona, Spain

Marta Mon
Instituto de Tecnología Química (UPV–CSIC), Universitat Politècnica de València–Consejo Superior de Investigaciones Científicas Avda. de los Naranjos s/n, Valencia, Spain

Oscar Pàmies
Universitat Rovira i Virgili, Departament de Química Física i Inorgànica, C/Marcel·lí Domingo, Tarragona, Spain

Preface

Catalysis is more alive and relevant than ever in trying to address the present and future challenges facing modern societies. Many research areas and industries depend on catalysts to construct molecules that can inhibit the progression of diseases, form elastic and durable materials, or store energy in batteries, among many other applications. Catalysis is present in a variety of fields, ranging from bio-, inorganic, organometallic, organic, solid-state, and theoretical chemistries. Interconnections between these fields are many times required for the design of an effective catalyst. Many research groups are working in two or more of these fields using approaches that span the areas of homogeneous, heterogeneous, and biocatalysis with applications ranging from the sustainable construction of molecules to the development of environmentally friendly technologies and in the advancement of new ways to generate clean energy, with remarkable results.

Following these considerations, Volume 69 of *Advances in Catalysis* contains four top quality chapters written by experts in different subfields of catalysis with the common aim of contributing to the *sustainable chemical production*. In Chapter 1, J.-E. Bäckvall and coworkers provide a comprehensive view of the use of metal–catalyzed aerobic oxidation of relevant organic substrates (alcohols, amines, alkenes, etc.). These oxidations that mimic biological systems are inspired either by the electron transfer respiratory chain or by natural metalloenzymes and allow the use of oxygen as a terminal environmentally friendly oxidant. In Chapter 2, A. Leyva-Pérez and coworker explain the advances over the past 10 years in the use of zeolites as catalysts for selective transformations of relatively large, functionalized molecules, which are used in advanced organic synthesis, fine chemistry, and pharma. The authors show that some organic reactions that until now had been dealt with homogeneous catalysts can also be carried out using zeolite-based catalysts, which are less toxic, cheaper, and recyclable, thus paving the way for their use in flow processes. In Chapter 3, O. Pàmies and coworker compile the latest advances in the synthesis of chiral carbo- and heterocyclic compounds via asymmetric metal-cycloaddition reactions with interceptive allylic substitution. In these reactions, a metal zwitterionic species reacts with a dipolarophile to form highly functionalized cyclic skeletons. This methodology is an appealing alternative to classical cycloadditions such as the Diels-Alder reactions, which are mainly governed by orbital symmetry

ix

considerations. Finally, in Chapter 4, M. Diéguez and coworkers collect the progress in the metal-catalyzed asymmetric hydroformylation of the less studied 1,1'-disubstituted alkenes as an atom-economical way to use readily available feedstocks to form chiral aldehydes that can be easily converted to other valuable building blocks.

We thank Mr. Sam Mahfoudh, who helped us initiate this project, and Ms. Tara A. Nadera and Mr. Jhon Michael Peñano whose help during the editing and production process was invaluable. We thank the Spanish Ministry of Science and Innovation (PID2019-104904GB-I00), the Catalan Government (2014SGR670), and the ICREA Foundation (ICREA Academia award to Montserrat Diéguez). Finally, our special thanks goes to all authors who contributed to this project.

MONTSERRAT DIÉGUEZ
OSCAR PÀMIES

> CHAPTER ONE

Metal-catalyzed biomimetic aerobic oxidation of organic substrates

Srimanta Manna[†], Wei-Jun Kong[†], and Jan-E. Bäckvall*

Department of Organic Chemistry, Arrhenius Laboratory, Stockholm University, Stockholm, Sweden
*Corresponding author: e-mail address: jeb@organ.su.se

Contents

1. Introduction		2
2. Metal-catalyzed biomimetic aerobic oxidations by multistep electron transfer		5
	2.1 Palladium-catalyzed biomimetic oxidation	5
	2.2 Ruthenium-catalyzed biomimetic oxidation	29
	2.3 Iron-catalyzed biomimetic oxidations with ETMs	36
3. Aerobic oxidation by biomimetic metal complexes		41
	3.1 Iron-catalyzed biomimetic oxidation	41
	3.2 Manganese-catalyzed biomimetic oxidation	44
	3.3 Copper-catalyzed biomimetic oxidation	46
	3.4 Osmium-catalyzed biomimetic oxidation	49
	3.5 Vanadium-catalyzed biomimetic oxidation	49
4. Concluding remarks and future perspectives		51
Acknowledgments		52
References		52
About the authors		56

Abstract

The use of molecular oxygen (O_2) as terminal oxidant in transition metal-catalyzed oxidative reactions is an appealing and challenging approach in organic chemistry. In these oxidations, the reoxidation of the reduced form of the metal catalyst by O_2 is the key and challenging step, due to the triplet ground state of O_2 and the high energy barrier for electron transfer. Inspired by (*i*) the electron transfer chain (ETC) in aerobic respiration and (*ii*) metalloenzymes for oxidations in Nature, organic chemists have developed metal-catalyzed biomimetic oxidations. In this chapter, two biomimetic approaches for transition metal-catalyzed aerobic oxidations are summarized, including: (*i*) biomimetic oxidation by multistep electron transfer pathways inspired by the respiratory

---

[†] These authors contributed equally to this work.

Advances in Catalysis, Volume 69
ISSN 0360-0564
https://doi.org/10.1016/bs.acat.2021.11.001

Copyright © 2021 Elsevier Inc.
All rights reserved.

chain and (*ii*) biomimetic oxidation by rational design of metal complexes that mimic metalloenzymes. The first type of aerobic oxidative transformations discussed include Wacker oxidations, alkyne oxidations, C—H functionalizations, and dehydrogenative oxidations of alcohols and amines. The second type of aerobic oxidations include iron-, copper- and manganese- osmium- and vanadium-catalyzed reactions.

1. Introduction

Oxidative reactions are the most common and fundamental transformations in organic chemistry, that have found broad applications in the synthesis of commodity chemicals and drug molecules *(1,2)*. However, expensive and waste-generating chemical oxidants, such as high valent metal salts (Cu^{II}, Ag^{I}, Cr^{VI} and Mn^{VII}) are employed in many cases, which severely compromise the economic and environmental merits of these reactions. Among various oxidants, molecular oxygen (O_2) is undoubtedly the most environmental-friendly and atom-efficient oxidant. O_2 has a strong oxidizing capability (+0.815 V vs HER in H_2O), which enables it to oxidize most organic compounds thermodynamically *(3,4)*. However, it is kinetically stable to reactions at ambient condition due to its triplet ground state *(5)*. The direct oxidations of organic compounds by O_2 is most of the time associated with low selectivity and harsh reaction conditions such as high temperature. Thus, the development of efficient, mild, and selective aerobic oxidation approaches has always been an important but challenging goal in organic chemistry.

Transition metal-catalyzed aerobic oxidative reactions, such as alcohol oxidations, C—H functionalizations and olefin functionalizations, etc., have become powerful tools in organic synthesis in the past decades. The usage of molecular oxygen as oxidant in these reactions only gives water as side product. In these oxidations the metal catalyst acts as a substrate-selective redox catalyst (SSRC), However, the reoxidation of the reduced form of the SSRC ($SSRC_{red}$) by O_2 is challenging due to the often high energy barrier for electron transfer from $SSRC_{red}$ to O_2 (Scheme 1A) *(6)*. In Nature, the electron transfer chain (ETC) in aerobic respiration is accomplished by a series of enzymes and coenzymes through multiple redox reactions (Scheme 1B) *(7–9)*. Specifically, the electrons in NADH have been transferred to O_2 by a series of enzymes including NADH dehydrogenases, cytochrome bc_1 complex and cytochrome c oxidase, while coenzymes ubiquinone (Q) and cytochrome c (Cyt c) play the role as electron transfer mediators (ETMs) between the enzymes. This strategy has largely lowered

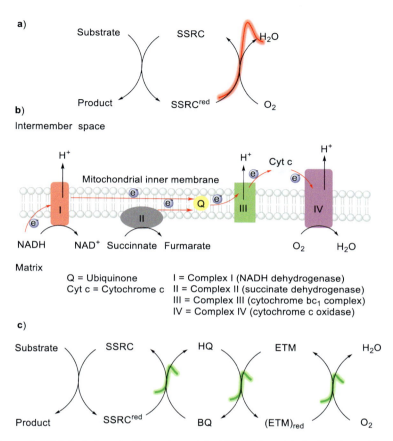

Scheme 1 (A) Substrate-selective redox catalyst (SSRC) in catalytic direct aerobic oxidation with high energy barrier; (B) The electron transfer chain (ETC) in aerobic respiration (respiratory chain); (C) Biomimetic aerobic oxidation approach with ETMs.

the overall energy barrier for the aerobic oxidation reaction. Inspired by the ETC in respiration, Bäckvall's group developed a biomimetic oxidation strategy in 1987, where a quinone and a metal-macrocyclic complex were used as ETMs (Scheme 1C) *(10)*. This biomimetic oxidation principle has been extended to other systems with different catalysts and transformations.

Nature has evolved a plethora of enzymes for the activation and utilization of triplet ground state O_2 with excellent efficiency and selectivity under mild reaction conditions. Among these oxygen-activating enzymes, monooxygenases are the most versatile ones that could directly activate O_2 and incorporate one oxygen atom into organic substrates *(11,12)*.

These oxygenation reactions typically include Baeyer-Villiger oxidation, epoxidation of alkene, C—H hydroxylation and heteroatom oxygenation. Metallic monooxygenases contain a metallic cofactor such as Fe, Cu, and Mn for the oxygen activation. Structurally, iron oxygenases are classified into heme dependent (also referred as Cytochromes P450) or nonheme enzymes such as Rieske dioxygenanase and αKG oxygenase (Scheme 2A). To mimic the enzymatic oxygenation, chemists have devoted much efforts to the development of macrocyclic metal complexes such as metalloporphyrin and tetradentate aminopyridine (N4) coordinated complex for oxygenation reactions (Scheme 2B). However, most of the developed biomimetic metal catalysts use hydrogen peroxide (H_2O_2) or other monooxygen donors (PhIO, alkyl hydroperoxides, mCPBA, etc.) as the terminal oxidation, which have been elegantly summarized in some recent reviews *(13–15)*. The development of biomimetic catalysts that are able to activate and use O_2 as terminal oxidant directly for oxygenation reaction is highly challenging *(16)*. The advances of this kind of biomimetic aerobic reactions catalyzed by metal complexes such as those in Scheme 2 will also be discussed in this chapter.

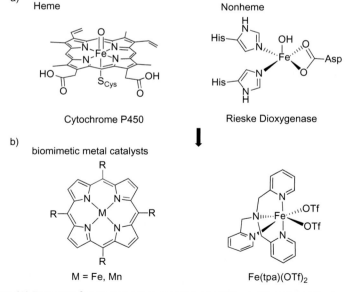

Scheme 2 (A) Enzymes for oxygenation reactions in Nature; (B) rational design of metal complexes mimicking metalloenzymes for oxygenation.

We aim to provide a comprehensive review on metal-catalyzed aerobic oxidations that mimic biological systems. This review will encompass the aforementioned two biomimetic strategies for metal-catalyzed aerobic oxidation of organic substrates: (i) biomimetic oxidation by multistep electron transfer pathways, which is inspired by the respiratory chain; and (ii) biomimetic oxidation by rational design of metal complexes that mimic metalloenzymes. The concept, mechanism and application of these biomimetic aerobic metal-catalyzed oxidations will be discussed. We hope this review will provide valuable insight into biomimetic oxidations for chemists who are interested in developing selective oxidation protocols using O_2 as the terminal oxidant.

2. Metal-catalyzed biomimetic aerobic oxidations by multistep electron transfer

2.1 Palladium-catalyzed biomimetic oxidation

Palladium-catalyzed oxidative reactions have been extensively studied and have become powerful tools in organic synthesis. The use of oxygen as terminal oxidant in palladium catalysis is appealing and still challenging. The formation of palladium black from the fast aggregation and precipitation of Pd^0 leads to deactivation of the catalyst and stops the reactions. The efficiency of reoxdiation of Pd^0 by oxygen constitutes the bottleneck for palladium-catalyzed aerobic oxidations. To solve this problem, efforts have been devoted to the development of air-stable ligands such as sulfoxide, pyridine and carbene to stabilize Pd^0 to accelerate its direct oxidation by oxygen. Advances in this aspect have been elegantly reviewed by Stahl and co-workers (17). Another approach to solve the problem with reoxidation is to use ETMs to facilitate the flow of electrons from Pd^0 to molecular oxygen via multiple redox steps (Scheme 3). This biomimetic electron relay strategy has been proved to be efficient in various palladium-catalyzed aerobic reactions including Wacker oxidations, C—H functionalizations and carbocyclizations (18,19).

2.1.1 Palladium-catalyzed Wacker oxidation

The Wacker oxidation is an excellent example of the principle of biomimetic oxidation using ETMs (20). This industrial process transforms ethylene to acetaldehyde by palladium-catalyzed oxidation with molecular oxygen. Catalytic amounts of cupric chloride are added and serve as an ETM. The Pd redox reaction of the Wacker process can be traced back

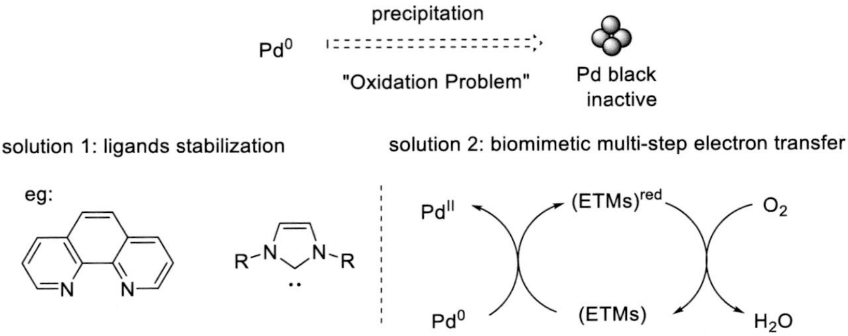

Scheme 3 Oxidation problem in Pd-catalyzed aerobic reactions and two different solutions.

to 1894 when Phillips observed that ethylene reacts with stoichiometric palladium chloride to give acetaldehyde, along with palladium black (21). In 1950s, researchers of Wacker Chemie greatly improved this reaction by using catalytic amounts of PdCl$_2$ and stoichiometric amounts of CuCl$_2$ under oxygen in acidic aqueous solution (22). Later on, a one-stage process was developed by Farbwerke Hoechst that employed catalytic amount of CuCl$_2$ (23). The key to success of the Wacker process is the use of CuCl$_2$ as an ETM that could easily transfer electrons from Pd^0 to oxygen (Scheme 4).

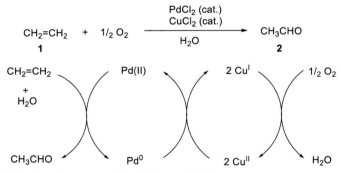

Scheme 4 Biomimetic approach in Wacker oxidation.

The mechanism of the Wacker oxidation concerning the stereochemistry of nucleophilic additions (syn/anti) to the olefin–palladium complexes has been studied substantially. An anti-hydroxypalladation process is strongly supported by experiments and computations (24).

The success of the Wacker process has spurred extensive research on palladium-catalyzed oxidative transformation of alkenes to carbonyl compounds, namely the Wacker–Tsuji reaction (25,26). In 1964, Selwitz found that dimethyl formamide (DMF) is an efficient co-solvent that addresses the insolubility problem and promotes the oxidation of higher order olefins *(27)*. Based on this observation, Tsuji has explored the selective oxidations of various functionalized terminal olefins and demonstrated its potential in complex molecule synthesis.

The presence of chloride anion was found to have a deleterious effect on the efficiency and selectivity in Wacker oxidation. The reaction rate is strongly inhibited by chloride ions and a high concentration of chloride leads to the formation of side product chlorohydrin. Thus, the development of a chloride-free Wacker oxidation is desirable. Bäckvall developed a biomimetic process for the oxidation of alkenes using 1,4-hydroquone (HQ) and a macrocyclic metal complex, such as cobalt Schiff base complexes (e.g. Co(salophen)) or iron phthalocyanine (FePc) as ETMs (Scheme 5) *(28,29)*. This chloride-free Wacker oxidation system showed high efficiency in terms of yield of product and was 16 times faster than the chloride-based oxidation process.

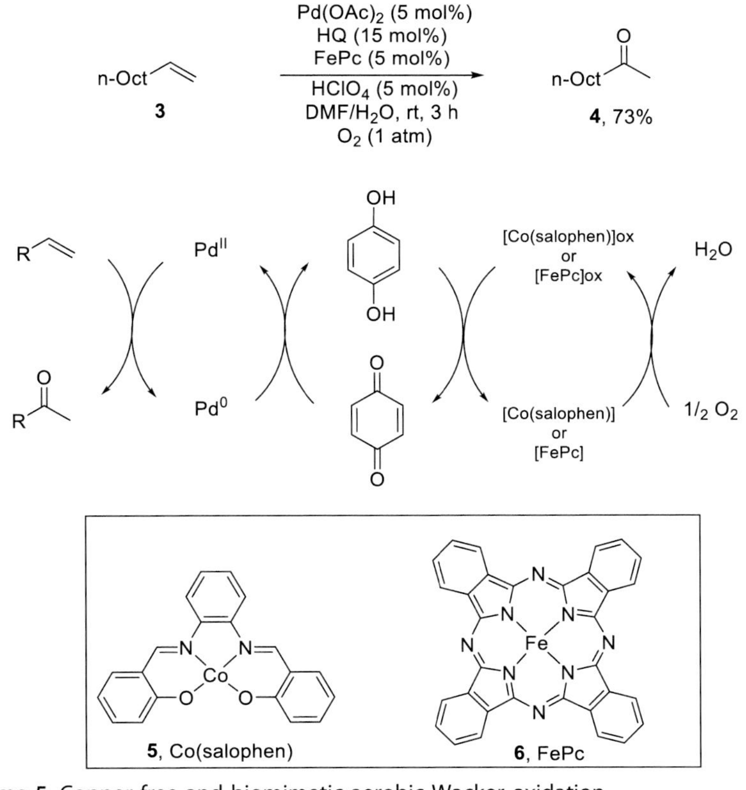

Scheme 5 Copper free and biomimetic aerobic Wacker oxidation.

Polyoxometalates containing redox reactive transition metals such as vanadium (V) have shown the ability to activate oxygen and has played the role of ETMs between palladium and molecular oxygen *(30)*. An alternative industrial process using phosphomolybdovanadate instead of CuCl$_2$ for palladium-catalyzed aerobic oxidation of alkene was developed by Catalytica. The concentrates of palladium and chloride are reduced to about 1/100th of those in the Wacker system (Scheme 6). In this catalytic system, vanadium(V) in the polyoxometalate functions as the ETM. The polyoxometalate matrix could maintain a high concentration of vanadium ions in solution and facilitate the rapid oxidation of vanadium (IV) by oxygen. While the simple hydrolytic vanadium (IV) specie VO^{2+} shows low solubility in the acidic solution and cannot be efficiently oxidized by oxygen.

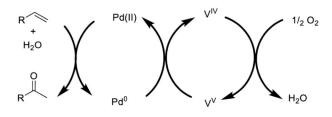

Scheme 6 Wacker oxidation with phosphomolybdovanadate as ETM.

When terminal alkenes are applied, the Wacker oxidation generally delivers methyl ketone as major product because of the Markovnikov selectivity in the hydroxypalladation step. However, this selectivity could be altered by the steric or chelation property of substrate and the addition of auxiliary ligands (Scheme 7) *(31)*. For example, styrene *(32)*, phthalimide protected allyl amine *(33)* and 3-vinyl isobenzofuranone *(34)* delivered the corresponding anti-Markovinkov products under the typical Tsuji-Wacker condition.

Nitrite salts were found to change the regioselectivity of the Wacker oxidation. In 1986, Feringa reported that the nitrite anion coordinated palladium catalyst could oxidize 1-decene to aldehyde and ketone in a 3:2 ratio under aerobic condition. The addition of KNO$_2$ further improved the selectivity of aldehyde selectivity to 7:3 (Scheme 8). From the proposed mechanism, nitrite anion serves as the role of ETM, as well as oxygen atom transfer mediator *(35)*.

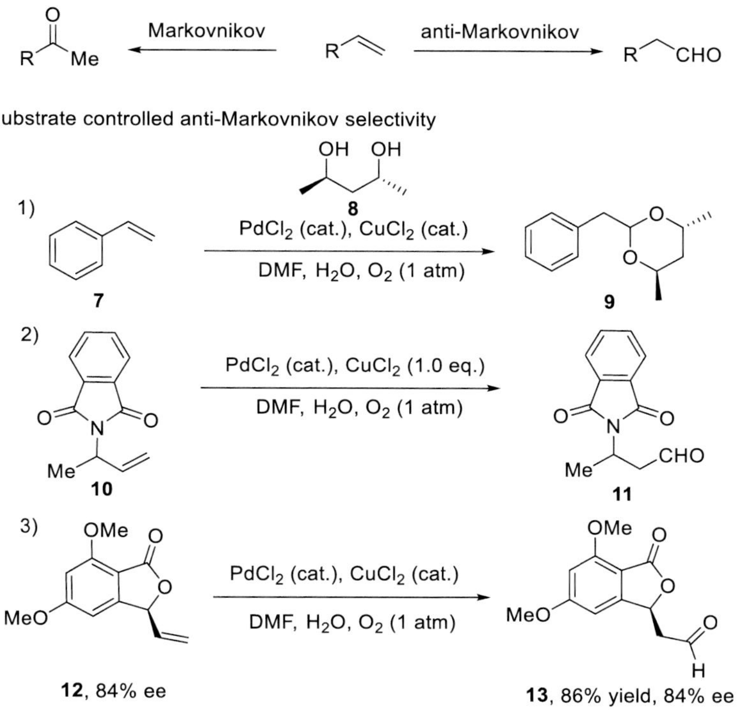

Scheme 7 Selected Wacker-Tsuji reactions with anti-Markovinkov selectivity.

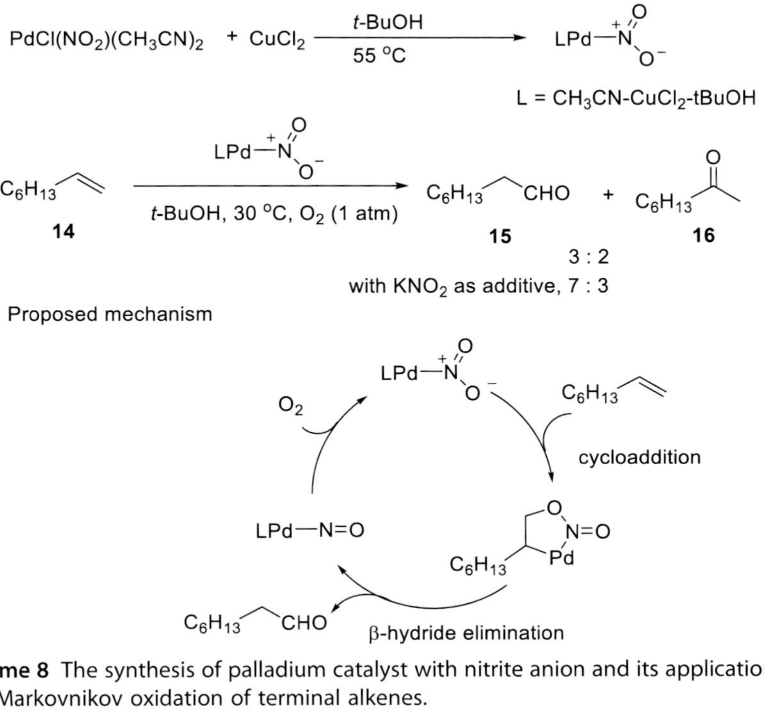

Scheme 8 The synthesis of palladium catalyst with nitrite anion and its application for anti-Markovnikov oxidation of terminal alkenes.

Recently, Grubbs and co-workers have established robust reaction conditions for anti-Markovnikov oxidation of stereoelectronically unbiased terminal olefins (Scheme 9) *(36,37)*. A series of linear α-olefins and α-olefins bearing functional groups such as carboxylic acid, halide and ester delivered aldehyde products in satisfactory yields and selectivity. The typical reaction condition contains $PdCl_2(PhCN)_2$, $CuCl_2$ and $AgNO_2$ in a co-solvent of *tert*-butanol/$MeNO_2$ at 20–25 °C. The addition of $AgNO_2$ was the key to the anti-Markovnikov selectivity. The authors speculated that the reaction system might facilitate the formation of an NO_2 radical, which would favor a radical type addition to the terminal position of the olefin (Scheme 9). However, the detailed mechanism for the anti-Markovnikov selectivity remains unclear and no evidence for an NO_2 radical intermediate has been provided.

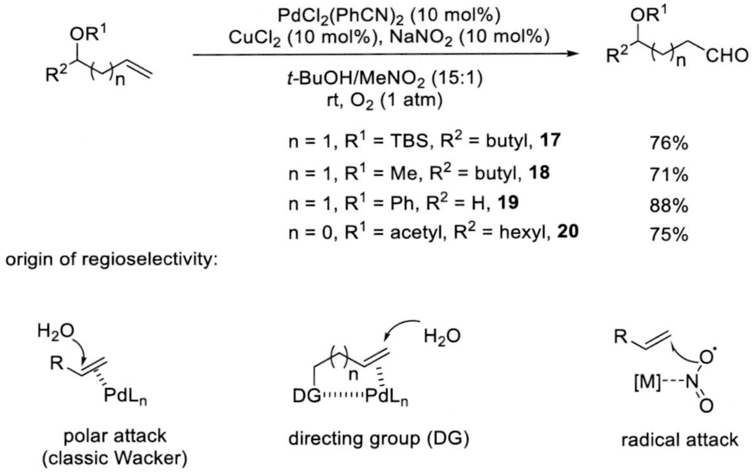

n = 1, R^1 = TBS, R^2 = butyl, **17** — 76%
n = 1, R^1 = Me, R^2 = butyl, **18** — 71%
n = 1, R^1 = Ph, R^2 = H, **19** — 88%
n = 0, R^1 = acetyl, R^2 = hexyl, **20** — 75%

Scheme 9 A general method for anti-Markovnikov oxidation of terminal alkenes and the origin of regioselectivity.

The Wacker oxidation system could be extended to nucleophiles other than water. For example, Stahl has developed conditions for divergent oxidation of styrene with carbamate (Scheme 10) *(38)*. The anti-Markovnikov product **22** was formed with $PdCl_2(CH_3CN)_2$, while $PdCl_2(NEt_3)_2$ delivered the 1,1-substituted enamine **23**.

The oxidation of internal olefins to ketones is quite challenging due to their low reactivity and selectivity compared to terminal ones. Kaneda has

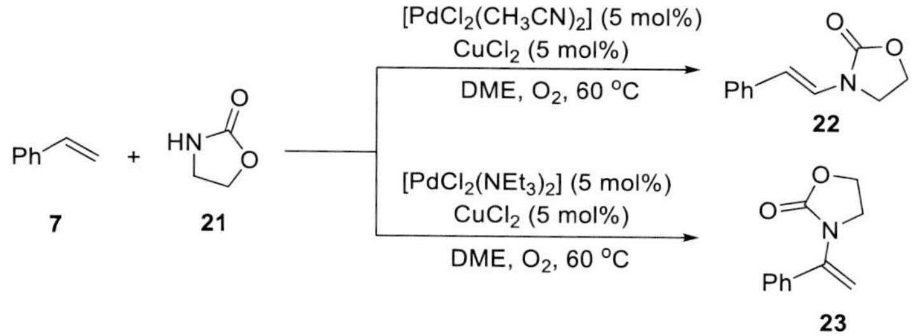

Metal-catalyzed biomimetic aerobic oxidation

Scheme 10 Regioselective Pd-catalyzed Wacker oxidation with nitrogen nucleophile.

reported a copper-free system for the oxidation of internal olefins (Scheme 11A) *(39)*. However, pressured oxygen in an autoclave is required. In 2013, Grubbs has developed an efficient and practical system for the oxidation of internal olefins, where BQ and iron phthalocyanine (FePc) served as the ETMs (Scheme 11B) *(40)*. It is worth noting that tetrafluoroboric acid (HBF_4) is essential for the formation of a plausibly reactive cationic complex intermediate. This reaction demonstrated excellent functional group tolerance.

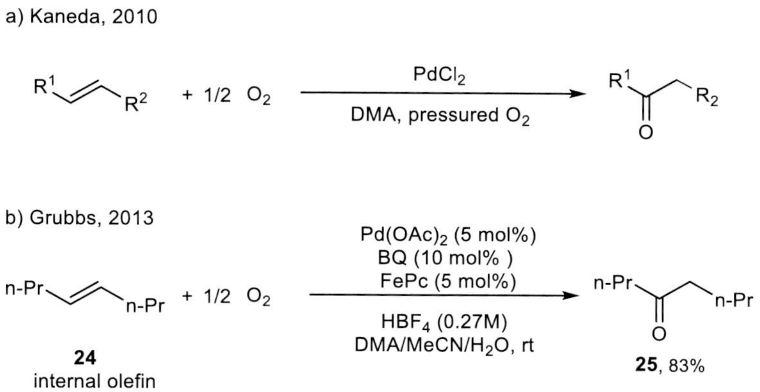

Scheme 11 Pd-catalyzed aerobic oxidation of internal alkenes. (A) Kenada in 2010; (B) Grubbs in 2013.

The biomimetic aerobic oxidation developed by the Bäckvall group was first applied to the 1,4-difuncionalization of 1,3-dienes in 1987 (Scheme 12A) *(10)*. Macrocyclic metal complex Co(TPP) (cobalt *meso*-tetraphenylporphyrin) demonstrated the fast and efficient electron transfer to oxygen in the palladium–hydroquinone–catalyzed oxidation of the diene. In 1990, the original system was extended to the use of the more practical

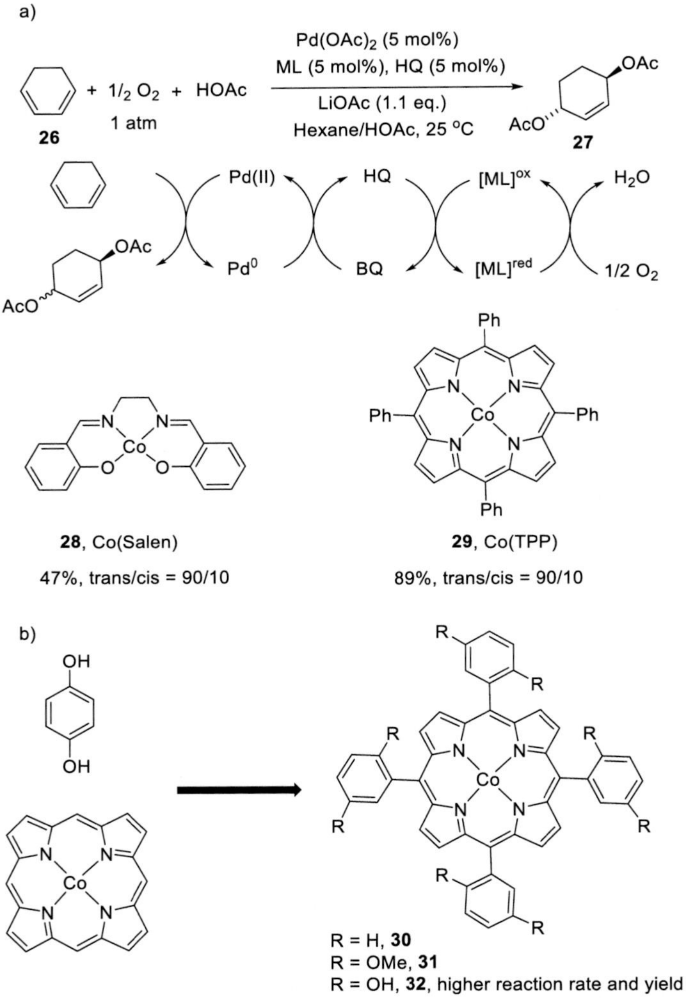

Scheme 12 (A) The development of biomimetic approach for palladium catalyzed aerobic oxidation of 1,4-dienes; (B) Integration of BQ and cobalt complex into one unit.

Co(salophen) as ETM together with the quinone *(29)*. Later on, a hybrid ETM, namely cobalt tetra(hydroquinone)porphyrin, was developed by integration quinone and metal porphyrin into one unit (Scheme 12B) *(41)*. Fast intramolecular electron transfer from the hydroquinone part to the oxidized cobalt center via the π system of the porphyrin increased the efficiency of the aerobic oxidation.

The Wacker reaction system is not just limited to oxidation of alkenes to form carbonyl compounds. Other transformations of olefins to complex molecules are realized with this biomimetic strategy as well. Taking advantage of the Pd–*t*BuONO co-catalyst system, Kang and coworkers has achieved the oxidative cyclization of 2-vinylaniline to give indole derivatives with oxygen as terminal oxidant (Scheme 13). The same reaction condition led to the cycloisomerization of o–allylanilines and indole derivatives were formed as well. In the latter case, the allylaniline first isomerized to propenyl aniline isomer via palladium-catalyzed allylic C—H activation. The isomer formed underwent a Wacker oxidation followed by the spontaneous annulation to hemiaminal with formation of indole product by loss of a molecule of H_2O. The potential of the practical use of reaction was demonstrated by a gram scale synthesis of indomethacin *(42,43)*.

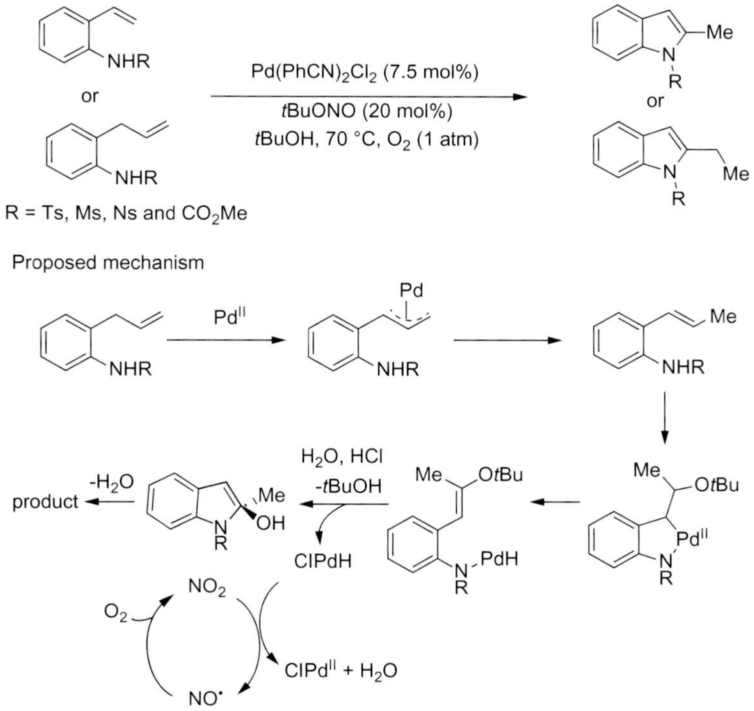

Scheme 13 Palladium-catalyzed oxidative cyclization of 2-vinylanilines and 2-allylanilines for the synthesis of indoles.

2.1.2 Palladium-catalyzed biomimetic oxidation of alkynes

The biomimetic multiple electron transfer oxidation strategy is also applicable for palladium-catalyzed transformations of alkynes. In 2009, Wan has reported the $PdBr_2/CuBr_2$-catalyzed aerobic oxidation of alkynes, which provides an easy access to 1,2-diketones (Scheme 14) *(44)*. The detailed mechanism of this transformation is not clear. However, control experiments showed that deoxybenzoin and benzoin are not the intermediates. Jiang and coworkers have developed a general protocol for the synthesis of 1-acetoxy-1,3-diene via an acetoxypalladation/Heck cross-coupling/β-H elimination tandem process (Scheme 15). The presence of $Cu(OAc)_2$ is crucial for the regeneration of Pd(II) and the resulting Cu(I) was oxidized to Cu(II) by O_2 *(45)*.

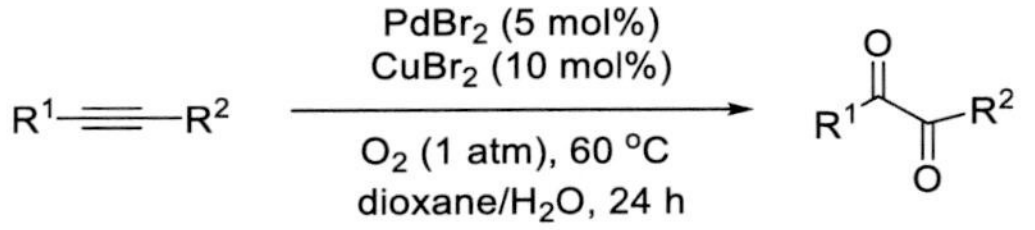

Scheme 14 $PdBr_2/CuBr_2$-catalyzed aerobic oxidation of alkynes to 1,2-diketones.

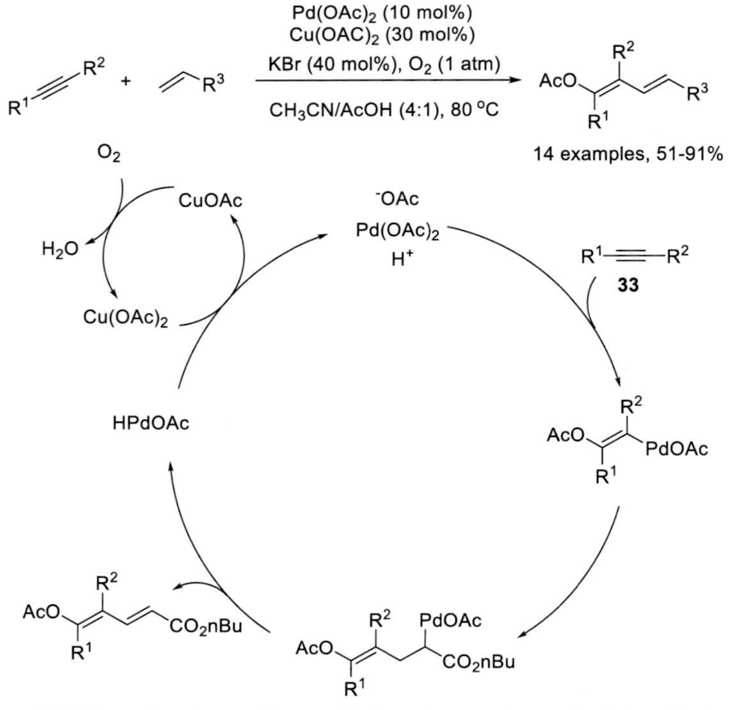

Scheme 15 Pd-catalyzed aerobic oxidation for acetoxypalladation/Heck cross-coupling/β-H elimination tandem process.

The same group also reported another elegant example of the palladium-catalyzed aerobic transformation of alkynes (Scheme 16) *(46)*. The [2+2+2] cycloaddition between 1,6-diynes and acrylates afforded polysubstituted aromatic cycles under typical Wacker oxidation condition with NMP as the solvent. Notably, a vinyl chloride intermediate was isolated and characterized, which provided mechanistic evidence for a chloropalladation of the alkyne as the initial step.

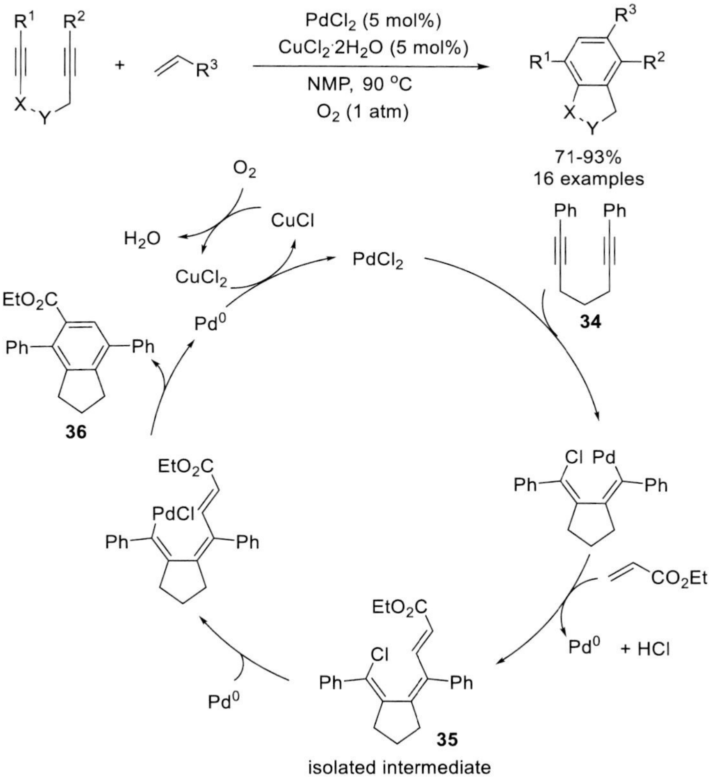

Scheme 16 Pd-catalyzed oxidative [2+2+2] cycloaddition between 1,6-diynes and acrylates.

2.1.3 *Palladium-catalyzed biomimetic C—H functionalization*
2.1.3.1 Palladium-catalyzed biomimetic allylic C—H functionalization
Palladium-catalyzed C—H activation has been extensively studied in the past two decades and become a powerful strategy in organic synthesis *(47,48)*. The use of oxygen as terminal oxidant in oxidative C—H functionalization is still an ideal and challenging goal. The biomimetic oxidation approach using ETMs has been explored in a variety of palladium-catalyzed

aerobic C—H functionalizations (Scheme 17). In 1990, the Bäckvall group reported the palladium-catalyzed aerobic allylic acetyloxylation of cyclohexene using HQ and FePc as ETMs (Scheme 17A) *(29)*. However, the generality in terms of alkenes of this reaction has not been explored and the nucleophile is limited to acetic acid. In 2010, Stambuli has developed a new thioether ligand for the palladium-catalyzed allylic acetoxylation of terminal olefins (Scheme 17B) *(49)*. The reaction delivered linear allylic acetates in good yields and regioselectivities. Catalytic amount of $Cu(OAc)_2$ was used under oxygen atmosphere to replace stoichiometric BQ. Still, only acetic acid was applied in this reaction. Recently, Stahl and co-workers have reported an allylic oxidation system that is compatible with diverse carboxylic acids (Scheme 17C) *(50)*. The regeneration of Pd(II) by aerobic oxidation of Pd^0 was enabled by the ligand 4,5-diazafluoren-9-one (DFA) and quinone/FePc co-catalyst.

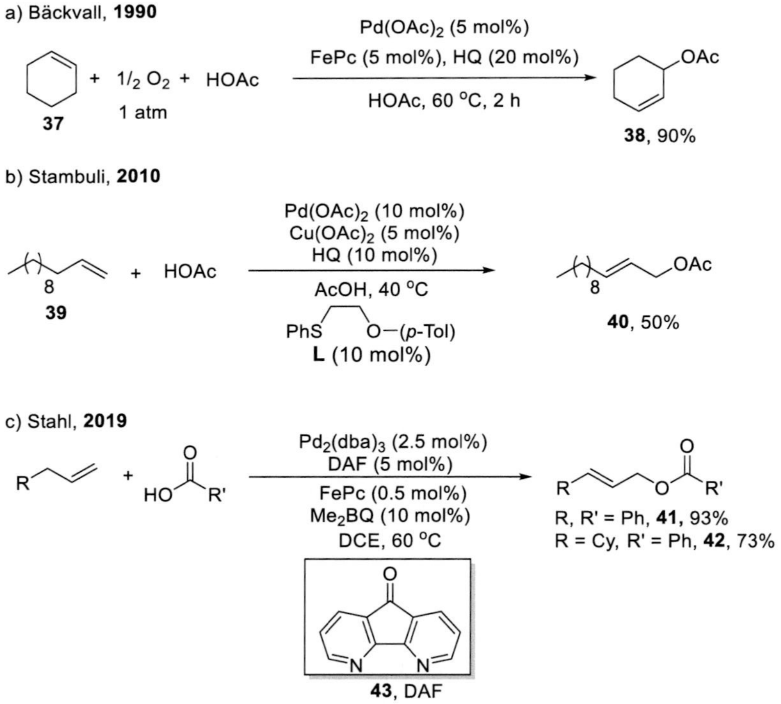

Scheme 17 Pd-catalyzed allylic C—H functionalizations with a biomimetic approach. (A) Palladium-catalyzed aerobic allylic acetoxylation of cyclohexene; (B) Palladium-catalyzed allylic acetoxylation of terminal olefins; (C) Palladium-catalyzed allylic oxidation with diverse carboxylic acids.

The advantage of the biomimetic oxidation strategy does not only lead to that expensive chemical oxidants are replaced by inexpensive molecular oxygen; other benefits might come along, such as increased reactivity. In 2016, White has reported the inhibition effect of high concentration of benzoquinone in palladium-catalyzed oxidative allylic C—H amination reaction (Scheme 18) *(51)*. The problem was solved by taking a biomimetic redox relay approach using molecular oxygen as terminal oxidant, in which catalytic amounts of cobalt salophen and hydroquinone were used as ETMs. Higher turnover numbers and product yields were obtained with the new system (Scheme 18).

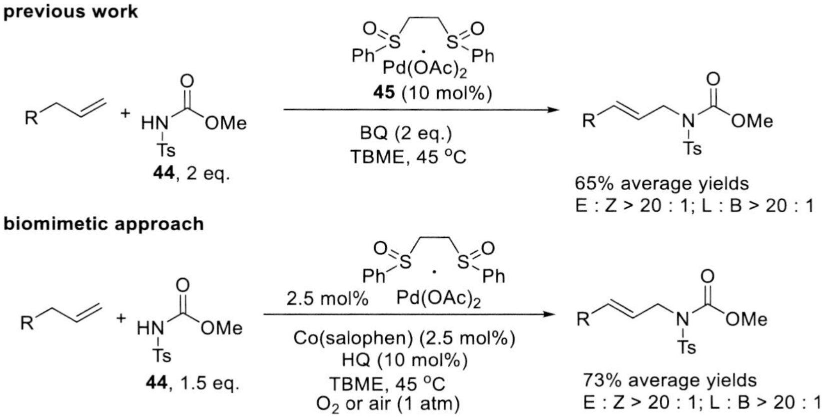

Scheme 18 The inhibition effect of high concentration of benzoquinone and the solution with a biomimetic approach.

Since the pioneering work of Fujiwara and Moritani on the oxidative Heck reaction *(52–54)*, palladium-catalyzed oxidative transformation of C(aryl)-H bonds has been extensively studied and established as a powerful tool for the synthesis of substituted or fused aromatic compounds. The utilization of O_2 as terminal oxidant greatly improved the sustainability of these oxidative reactions. The multiple-step electron transfer biomimetic strategy were applied in some cases.

2.1.3.2 Palladium-catalyzed biomimetic aryl and olefinic C—H functionalization

In 2007, DeBoef and co-workers has developed a biomimetic oxidation system for palladium-catalyzed dehydrogenative cross coupling of benzene and benzofuran (Scheme 19) *(55)*. With heteropolymolybdovanadic acid

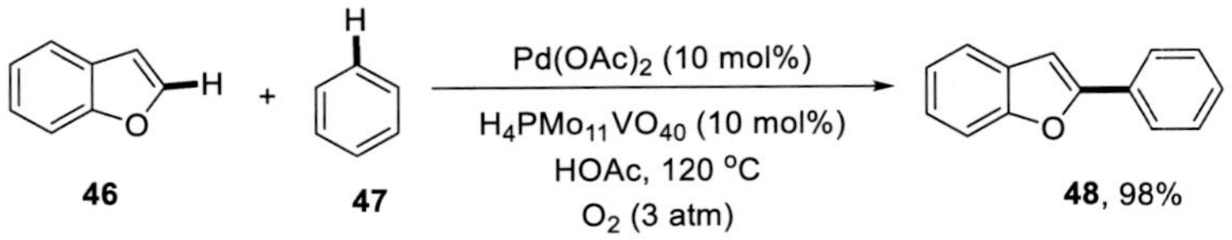

Scheme 19 Pd-catalyzed arylation of benzofuran with heteropolymolybdovanadic acid as ETM.

$H_4PMo_{11}VO_{40}$ (HPMV) as ETM, biaryls were synthesized with good to high yields in excellent regioselectivity under aerobic conditions, while other chemical oxidants such as AgOAc, $Cu(OAc)_2$, BQ and $PhI(OAc)_2$ gave low yields and poor selectivity between the C2 and C3 of the benzofuran. However, pressured oxygen (3 atm) was required and benzene was added as co-solvent in this reaction.

Bäckvall has realized oxidative dehydrogenative alkenylation of arenes with unbiased olefins using an efficient biomimetic aerobic strategy (Scheme 20) *(56,57)*. Catalytic amounts of Fe(Pc) and BQ were employed as electron-transfer mediators (ETMs) between palladium and 1 atm of oxygen. The reaction proceeds with good to excellent diastereoselectivity and regioselectivity.

Scheme 20 Pd-catalyzed biomimetic oxidative olefination of arenes.

The Stahl group described a highly efficient reaction system for the acetoxylation of benzene using fuming HNO_3 as redox mediator (Scheme 21) *(58)*. The loading of $Pd(OAc)_2$ could be reduced to 0.1 mol % and a turnover number (TON) of 136 was achieved under ambient oxygen pressure. The reaction proceeds with high chemoselectivity (phenyl acetate/nitrobenzene = 40:1) and no biphenyl product was observed.

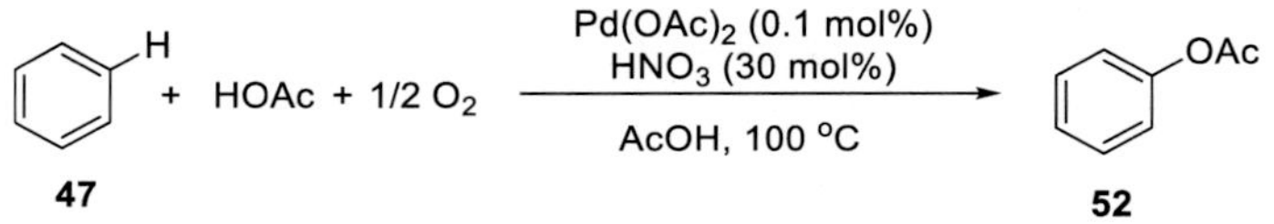

Scheme 21 Pd-catalyzed acetoxylation of benzene with nitric acid as ETM.

Jiang and Li developed a biomimetic palladium-catalyzed aerobic C—H activation cascade reaction in 2017 (Scheme 22) *(59)*. A series of dihydrobenzofurans were synthesized under ambient pressured oxygen (1 atm). Catalytic amount of $Cu(OAc)_2$ was used as the ETM to ensure efficient regeneration of (Pd^{II}) by O_2. The addition of BQ could increase the yield slightly, presumably due to the stabilization of Pd^0 through coordination. Mechanistically, the reaction proceeds through a series of tandem steps including olefin directed C—H activation, intramolecular olefin insertion, another insertion of activated olefins and β-hydride elimination. The resulting Pd^0 was reoxidized by the CuX_2/O_2 system.

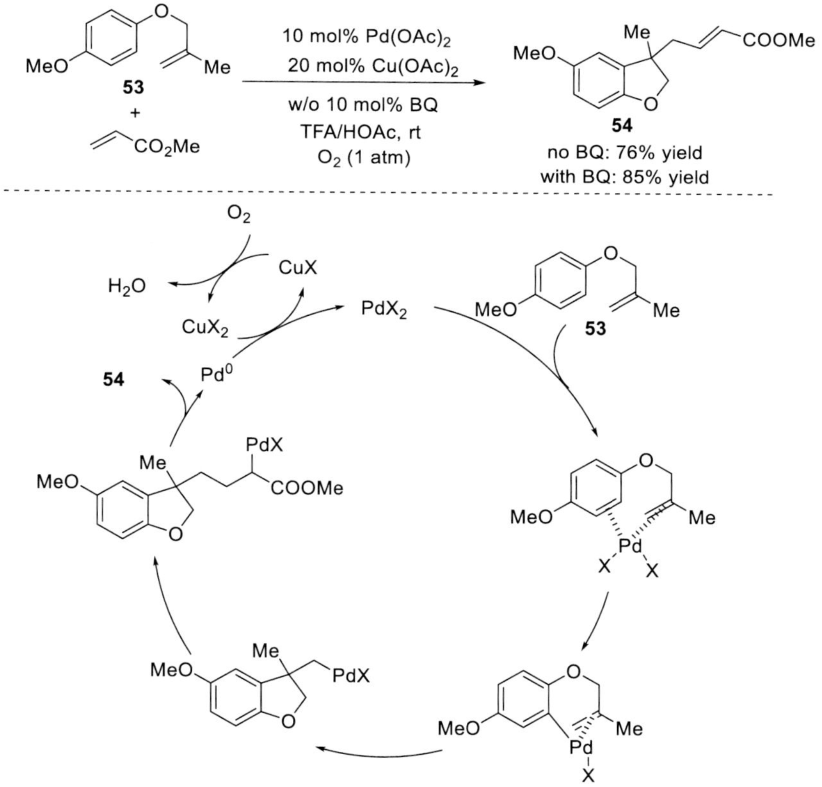

Scheme 22 Pd-catalyzed cascade C—H activation/olefination of phenol derivatives.

In 2014, Shi reported a palladium-catalyzed aerobic oxidative Heck reaction of phenethylamine using a removable 1,2,3-triazole directing group (Scheme 23A) *(60)*. $Cu(OTf)_2$ was identified as an efficient ETM, while

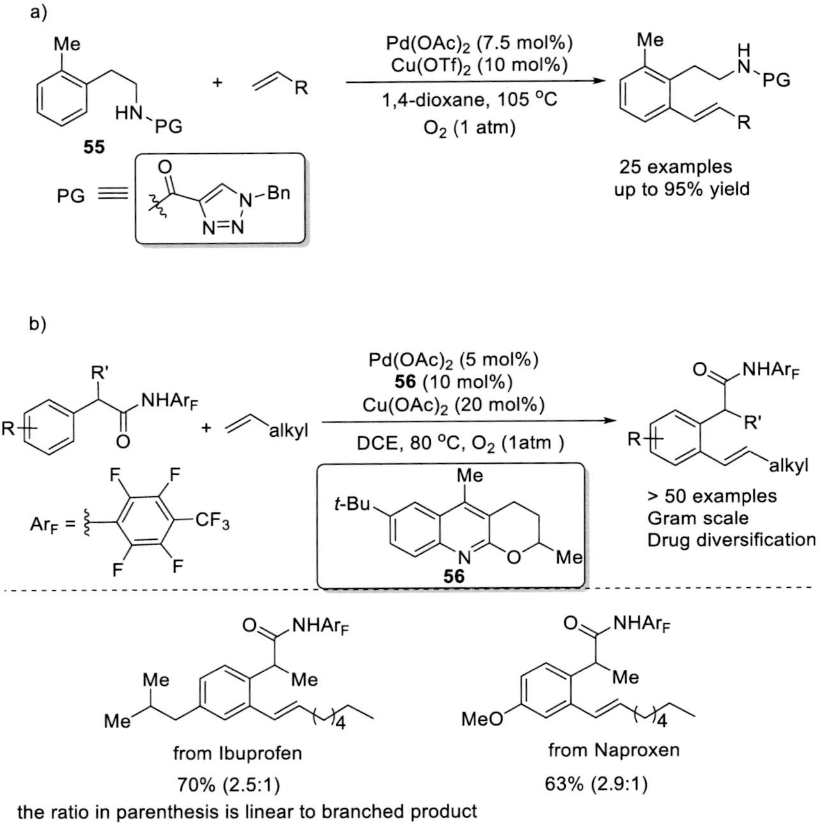

Scheme 23

Scheme 23 (A) Pd-catalyzed aerobic oxidative C—H olefination of phenethylamines; (B) Pd-catalyzed aerobic oxidative C—H olefination of phenylacetic acids.

$CuCl_2$ only gave trace amount of product. The group of Yu has reported the aerobic palladium-catalyzed *ortho*-olefination of phenylacetic acid derivatives using a similar strategy (Scheme 23B). In this case, $Cu(OAc)_2$ was used as the optimal ETM. A quinoline ligand was essential for the C—H activation step *(61)*. Moderate to good ratios of linear to branched product and excellent diastereoselectivity were achieved with a variety of electronically unbiased olefins. Two representative drug molecules ibuprofen and naproxen were functionalized successfully.

Transition-metal-catalyzed C—H functionalization of alkenes could provide expedient access to structurally diversified compounds, such as heterocycles, multi-substituted alkenes and 1,3-dienes *(62)*. However, it has not been thoroughly explored compared to C(alkyl)—H and C(aryl)—H activation, which might be due to its intrinsic high reactivity that influence the regioselectivity and diastereoselectivity. Engle and co-workers have developed a catalytic method to prepare highly substituted 1,3-dienes from alkenes through palladium(II)-catalyzed C(alkenyl)—H activation strategy using 8-amino quinoline amide (AQ) as directing group (Scheme 24) *(63)*. Aerobic oxidation conditions were made possible by the use of $Co(OAc)_2$ as ETM under oxygen.

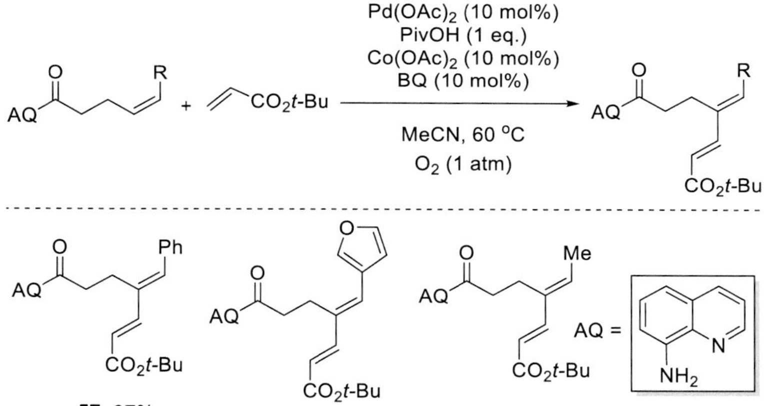

Scheme 24 Biomimetic oxidation approach for Pd-catalyzed directed C(alkenyl)—H olefination.

2.1.3.3 Palladium-catalyzed biomimetic alkyl C—H functionalization

An elegant example of the biomimetic strategy applied in palladium catalyzed $C(sp^3)$-H activation/functionalization was reported by Sanford in 2012 (Scheme 25) *(64)*. With $NaNO_3$ or $NaNO_2$ serving as ETM, the palladium–catalyzed reaction went smoothly under molecular oxygen. A variety of substrates bearing strong coordination directing groups such as oxime ether and quinoline were applicable and delivered the corresponding mono-, di- or tri-acetoxylated products. The NO_2 species could effectively oxidize Pd^{II} to Pd^{IV} and promote its reductive elimination, a catalysis regime that is different from aforementioned Pd^{II}/Pd^0 ones. The envisioned NO was detected by treatment with butylated hydroxytoluene (BHT), which afforded 2,6-di-*tert*-butyl-4-methyl-4-nitrosocyclohexa-2,5-dienone (TBMND).

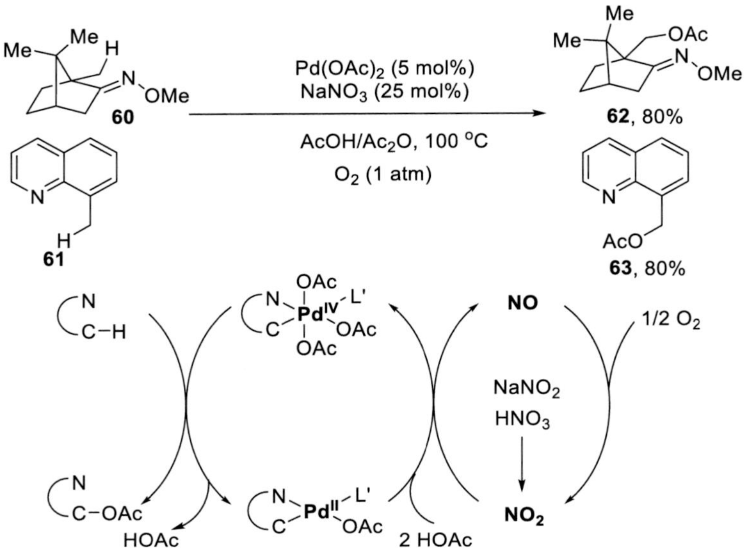

Scheme 25 Pd-catalyzed biomimetic aerobic C(sp³)–H acetoxylation.

2.1.3.4 Palladium-catalyzed biomimetic C—H functionalization of allenes

Allenes have always been fascinating compounds in organic chemistry because of their extraordinary structure and reactivity properties *(65,66)*. The group of Bäckvall has developed a series of palladium-catalyzed C—H activation/transformations of allenes, among which biomimetic aerobic oxidation strategy were widely applied to improve the efficiency and sustainability of the reactions. In 2006, the group developed an aerobic carbocyclization of cyclohexene-substituted allenes to form bicyclic compounds (Scheme 26A) *(67)*. FePc and BQ were applied as ETMs for this aerobic oxidative reaction. As similar approach was used for the carbocyclization of cyclohexadiene substituted allenes (Scheme 26B) *(68)*. Water was used as an effective nucleophile for the final allylic substitution step. An *exo* attack by water on the π-(allyl)palladium intermediate provides the stereospecificity of the reaction.

In 2013, Bäckvall and co-workers reported the oxidative cyclization of allenyne to multi-substituted cyclopentene by palladium-catalyzed aerobic biomimetic oxidation (Scheme 27) *(69)*. Mechanistically, it was proposed that the nucleophilic attack of the allene moiety on the palladium(II) delivers a dienylpalladium species. Subsequent intramolecular cyclization via alkyne insertion gives a vinyl palladium intermediate, followed by the coordination

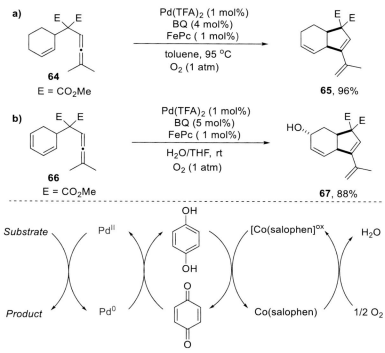

Scheme 26 Pd-catalyzed biomimetic oxidation of enallenes. (A) Cyclohexene substituted allenes; (B) Cyclohexadiene substituted allenes.

of terminal alkyne and reductive elimination to give the final product. The aerobic regeneration of Pd^{II} was accomplished by BQ/Co(salophen) catalysis system. Notably, the reaction proceeds with excellent chemoselecitivity and regioselectivity without the formation of dimerization of allenyne.

The palladium-catalyzed biomimetic oxidation approach was further applied for the carbocyclization of enallene (Scheme 28) *(70)*. A variety of substituted cyclohexenes were selectively synthesized. The coordination of the proximal olefin assists the allenic C—H activation and formation of a vinylpalladium intermediate *(71)*. A six-membered cyclohexane intermediate was formed via the insertion of the remote olefin. The transmetalation with B$_2$pin$_2$ or boronic acids, followed by reductive elimination afforded the desired product. The distance between the two olefin bonds is crucial for reactivity and regioselectivity. No side products, such as cyclobutene via the insertion of the proximal olefin, were formed in this reaction.

In previous biomimetic oxidation system with quinone and metal macrocycles as the ETMs, electron transfer proceeds mainly in an intermolecular

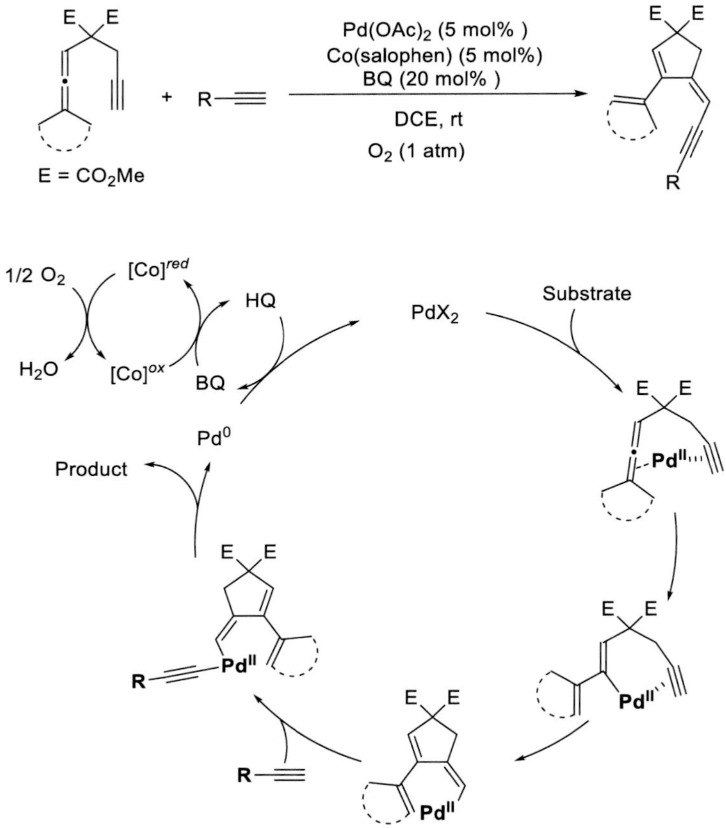

Scheme 27 Pd-catalyzed aerobic carbocyclizative alkynylation of allenynes.

fashion. The integration of these two ETMs into one unit might facilitate the electron transfer and improve the turnover frequency. In 1993, the Bäckvall group for the first time installed hydroquinone around the periphery of a cobalt prophyrin to afford a hybrid catalyst (Co(TQP), **Co-1**) that is efficient for Pd-catalyzed aerobic 1,4-diacetoxylation of cyclohexadiene *(41)*. However, this hybrid catalyst suffered from the drawback of tedious synthesis. A new type of hybrid ETM integrating hydroquinone and cobalt Schiff base complexes was developed by the same group in 2008 (Scheme 29) *(72)*. These readily accessible ETMs (**Co-2** Co(salmdpt)-HQ, and **Co-3** Co(salophen)-HQ) were shown to be highly efficient in the palladium-catalyzed aerobic oxidative carbocyclization of enallene (Scheme 29B) *(73)*. In 2010, the **Co-2** catalyst was used as an efficient ETM for the oxidative cyclization of aza-enallene **72** to form dihydropyrroles **73** *(74)*.

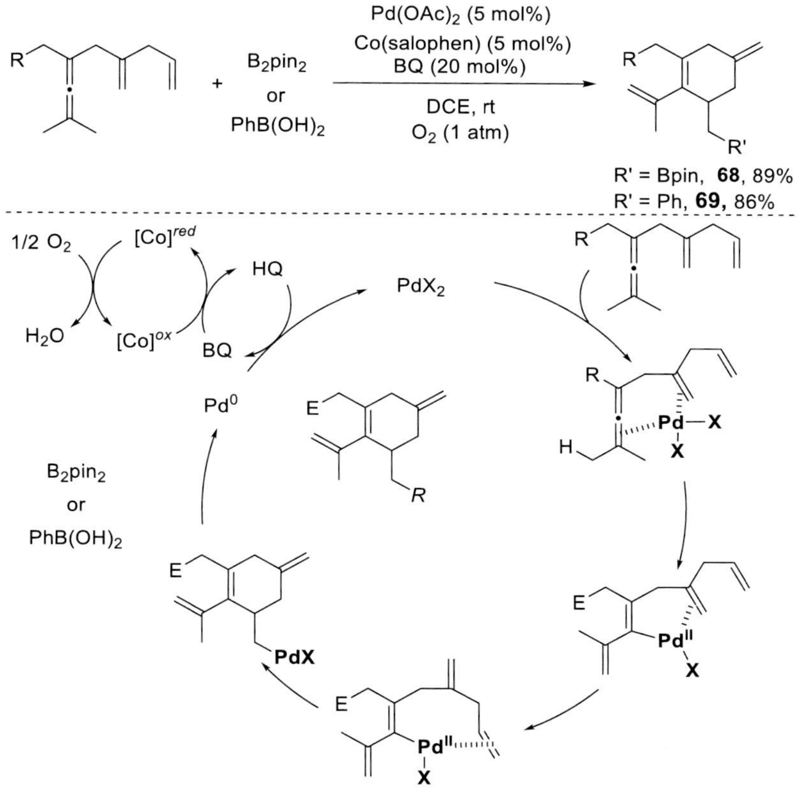

Scheme 28 Pd-catalyzed aerobic carbocyclization of dienallenes and proposed mechanism.

Through the exquisite design of substrates, Bäckvall and co-workers realized the palladium-catalyzed oxidative arylation of enallenyne in 2018 (Scheme 30A) *(75)*. Hybrid catalyst **Co-3** was identified as the optimal ETM for this aerobic oxidative carbocyclization reaction. The olefin proximal to the allene plays the role of directing group for the allenic C—H activation. Kinetic experiments demonstrated that the reaction with hybrid ETM proceed faster than those with separate ETMs. Various multi-substituted cyclohexenes were synthesis at room temperature under ambient pressured oxygen. The same strategy of aerobic oxidation was further applied to the carbocyclization of dienallene for the stereoselective synthesis of *cis*-1,4-disubstituted six-membered heterocycles, including dihydrophyran and tetrahydropyridine derivatives (Scheme 30B) *(76)*. Experimental and computational studies showed that the pendant olefin is essential for the regioselective allenic C—H cleavage and the *cis* selectivity of the reaction.

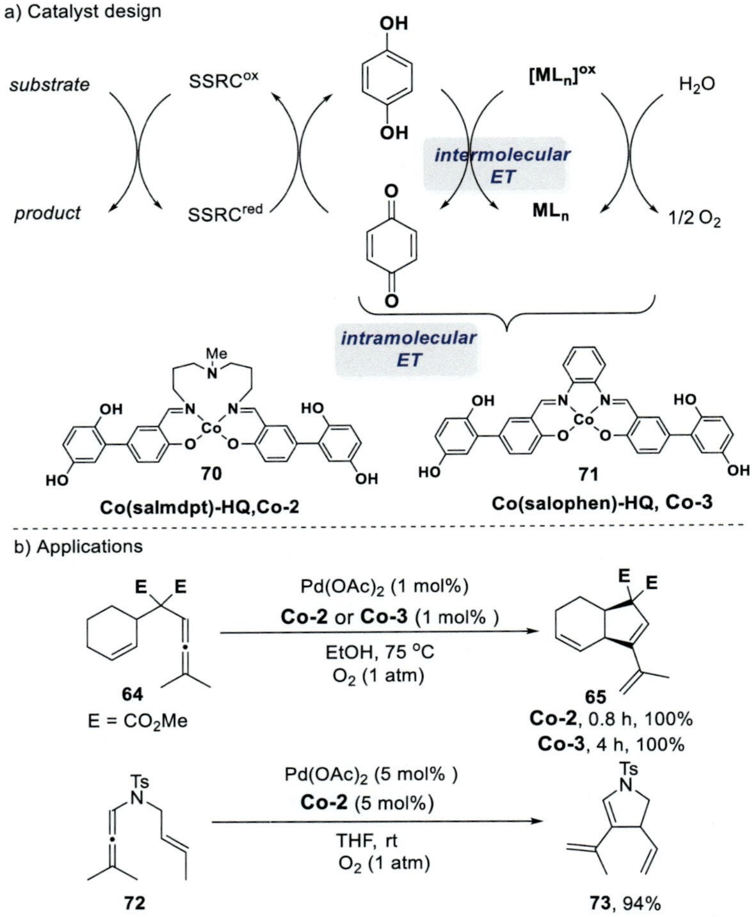

a) Catalyst design

b) Applications

Scheme 29 (A) Design of a bifunctional hybrid ETM for aerobic oxidations; (B) Applications in Pd-catalyzed carbocyclization of enallenes and aza-enallenes.

Seven-membered cyclic oxo-dienes were synthesized when the allylic group of **76** was replaced by an allenic substituent under the same reaction conditions (Scheme 30C) *(77)*.

2.1.4 Heterogeneous palladium-catalyzed biomimetic oxidation

Heterogeneous palladium catalysis has been extensively studied in cross-coupling reactions for the reason of high efficiency and recyclability *(78–81)*. However, it has been rarely explored in aerobic oxidative reactions. In 2019, Jiang developed an elegant heterogeneous Pd-catalyzed aerobic oxidation system (AOS) in which Pd^{II}, phenanthroline ligand, and Cu^{II}

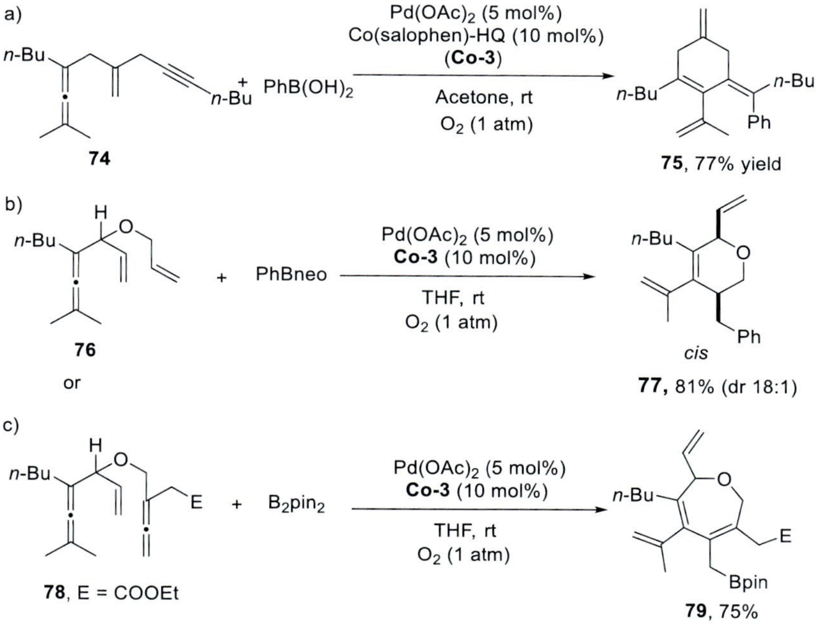

Metal-catalyzed biomimetic aerobic oxidation

Scheme 30 Pd-catalyzed aerobic carbocyclization with a hybrid ETM **Co-3**. (A) Arylation with enalleneynes; (B) Arylation with dienallenes; (C) Boronation with diallenes.

were immobilized onto a metal–organic framework (MOF, **82**) (Scheme 31) *(82)*. The proximity of palladium and copper facilitates the efficient electron transfer between them and increases the reaction turnover frequency. The MOF-based catalyst exhibited better performance than the homogeneous counterparts in terms of TONs.

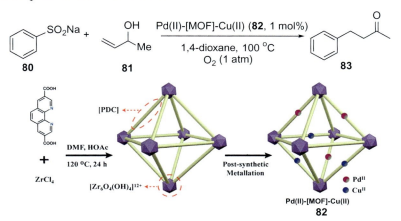

Scheme 31 Metal-organic framework (MOF)-based catalyst for aerobic oxidation (Ref. *(82)*). Copyright Wiley-VCH GmbH. Reproduced with permission.

It is generally considered that hydroquinone (HQ) cannot be oxidized directly by molecular oxygen, unless the pH is high (pH > 13). Therefore, an oxygen activating ETM such as Fe(Pc) or Co(salophen) is generally required in the biomimetic oxidation strategy *(18,19)*. Very recently, the group of Bäckvall has revealed that hydroquinone can indeed be directly oxidized by molecular oxygen in a co-solvent of methanol and acetonitrile. Based on this observation, a heterogeneous palladium-catalyzed aerobic allenic C—H arylation was realized, and BQ served as the only ETM between Pd^0 and O$_2$ (Scheme 32). The reoxidation of HQ to BQ proceeds

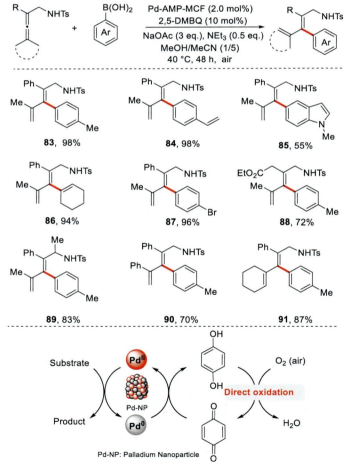

Scheme 32 Heterogeneous palladium-catalyzed aerobic allenic C—H arylation with BQ as the direct ETM.

at a moderate rate but fast enough to provide high yield of coupling product in the aerobic oxidation. Additionally, mechanistic studies show that BQ also plays the role as ligand in the reductive elimination step. Various multi-substituted 1,3-dienes (**83–91**) were synthesized under mild reaction conditions under air *(83)*. This new regime of biomimetic oxidation is believed to be applicable in other transition metal-catalyzed aerobic oxidation reactions.

2.2 Ruthenium-catalyzed biomimetic oxidation

Metal-catalyzed oxidation constitutes one of the pivotal transformations in organic synthesis and many important organic transformations involve oxidation steps *(84)*. In the past few decades Ru-catalyzed aerobic oxidation received much attention owing to the simplicity of the reaction protocols and broad functional group compatibilities *(85)*. A dimeric ruthenium complex called the Shvo catalyst, first synthesized in 1980s *(86)*, constitutes an important catalyst that has been used for highly efficient oxidation of alcohols *(87)*and amines *(88)* by the Bäckvall's group. Thermal dissociation of dimeric ruthenium complex **Ru-1** provides the monomers **Ru-2** and **Ru-3**, which are considered to be the active catalyst (Scheme 33). The cyclopentadienone in **Ru-3** is proposed to be the active dehydrogenation catalysts in the reaction of alcohols and amines. On the other hand, the hydroxycyclopentadienyl complex **Ru-2** has a proton on the oxygen of the cyclopentadienyl ligand and a hydride on the Ru center, and is considered to be the active hydrogenation catalyst. The proton and the hydride are involved in hydrogen transfer reactions. The cyclopentadienone in **Ru-3** has a proton acceptor site as well as a hydride-accepting Ru center and can catalyze oxidations. Recently, Bäckvall's group extensively used these kind of ruthenium complexes for the aerobic biomimetic oxidation of alcohols *(87)* and amines *(88)*. In the following section Ru-catalyzed aerobic

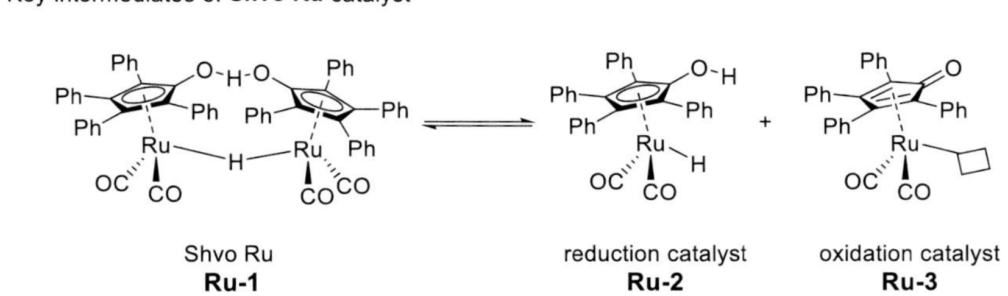

Key intermediates of **Shvo Ru** catalyst

| Shvo Ru | reduction catalyst | oxidation catalyst |
| **Ru-1** | **Ru-2** | **Ru-3** |

Scheme 33 Shvo's ruthenium complexes

biomimetic oxidation of organic substrates will be outlined and different strategies using multistep electron transfer will be discussed.

2.2.1 Ruthenium-catalyzed alcohol oxidation

The oxidation of alcohols is an important transformation in organic chemistry that is involved in many biological reactions *(89)*. Oxidation technologies that are inspired by biological systems are becoming increasingly popular *(18)*. In 1994, Bäckvall and co-workers reported the first efficient Ru-catalyzed biomimetic oxidation of alcohols as a complementary route to prepare ketones employing 2,6-di-*tert*-butyl-1,4-benzoquinone **94a** and Co(salmdpt) complex **Co-4** as electron transfer mediators under molecular oxygen atmosphere (Scheme 34) *(87)*. The oxidation of alcohols could be performed selectively to obtain the corresponding carbonyl products in generally good yields. Later in 2002, the Bäckvall group extended Ru-catalyzed efficient biomimetic oxidation of various alcohols under

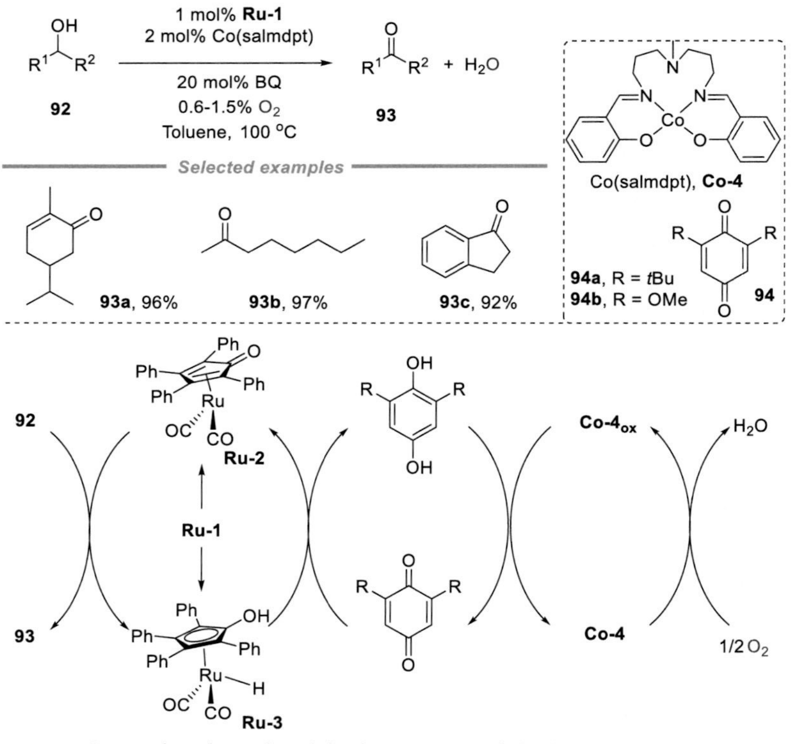

Scheme 34 Ru-catalyzed aerobic dehydrogenation of alcohols.

Metal-catalyzed biomimetic aerobic oxidation 31

molecular oxygen to access carbonyl compounds using 2,6-dimethoxy-1,4-benzoquinone **94b** and Co(salmdpt) complex **Co-4** as electron transfer mediators *(90)*. The electron-rich quinone (2,6-dimethoxy-1,4-benzoquinone) was shown to be the best quinone as an electron transfer mediator (ETM) in the catalytic process. It is most likely that the reoxidation of the more electron-rich 2,6-dimethoxy-1,4-hydroquinone is oxidized much faster than the 1,4-*di-tert*-butyl-hydroquinone by the **Co-4$_{ox}$**. The ruthenium hydride intermediate **Ru-3** is formed after reaction with alcohol **92**, which is oxidized to ketone **93**. An outer-sphere mechanism for the dehydrogenation of the alcohol by Shvo's complex was proposed. Initially the Shvo complex undergoes thermal dissociation to form two active catalysts **Ru-2** and **Ru-3**. Then the active ruthenium catalyst **Ru-2** reacts with alcohol to produce ketone and the ruthenium hydride complex **Ru-3**. The active ruthenium catalyst **Ru-2** and hydroquinone is regenerated after reaction with ruthenium hydride intermediate **Ru-3** and benzoquinone derivative. Afterwards, hydroquinone is oxidized to benzoquinone by an oxidized form of the cobalt macrocycle (**Co-4$_{ox}$**). The reduced form of the cobalt macrocycle (**Co-4**) reacts with molecular oxygen to give an oxidized cobalt macrocycle (**Co-4$_{ox}$**).

2.2.2 Ruthenium-catalyzed amine oxidation

Following on from these reports, the Bäckvall's group envisioned that the Ru-catalyzed biomimetic approach could be applied for an efficient dehydrogenation of amines. Subsequently, Bäckvall and co-workers developed a Ru-catalyzed efficient oxidation of secondary amines using a biomimetic approach to access imine derivatives in 2005 *(88)*. This method provided milder conditions for the aerobic oxidations of amines. A wide range of aliphatic and aromatic amines were oxidized to their corresponding imines in good yields (Scheme 35). It was found that electron-rich amines react much faster than electron-deficient amines. A similar electron transfer mechanism to that reported for the alcohol oxidation was proposed for the oxidation of amine. First, two active catalysts, **Ru-2** and **Ru-3**, are formed after thermal dissociation of Shvo complex **Ru-1**. Then the active ruthenium catalyst **Ru-2** reacts with amine to produce imine and the ruthenium hydride complex **Ru-2**. The active ruthenium catalyst **Ru-2** and hydroquinone is regenerated after reaction with ruthenium hydride intermediate **Ru-3** and benzoquinone. Afterwards, hydroquinone is oxidized to benzoquinone by an oxidized form of the cobalt macrocycle (**Co-4$_{ox}$**). In this process

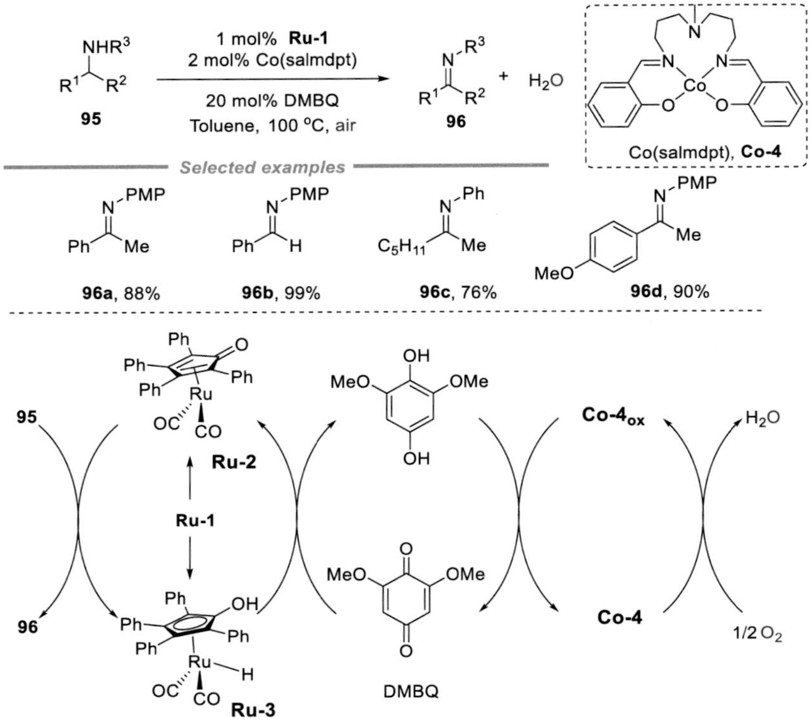

Scheme 35 Ru-catalyzed aerobic dehydrogenation of amines.

cobalt macrocycle **Co-4** is formed, which reacts with molecular oxygen to give the oxidized cobalt macrocycle (**Co-4$_{ox}$**) and H_2O.

An inner-sphere mechanism for the amine oxidation was proposed (Scheme 36) *(88)*. The Shvo catalyst can react with amine to form ruthenium–amine complex **Ru-5**, which is confirmed by mechanistic studies at low temperature *(91,92)*. In this mechanism, the complex **Ru-4** then furnish the intermediate **Ru-5** followed by β-hydride elimination from amine to deliver imine. An alternative mechanism for hydrogen transfer from amine to Ru was also discussed *(92)*.

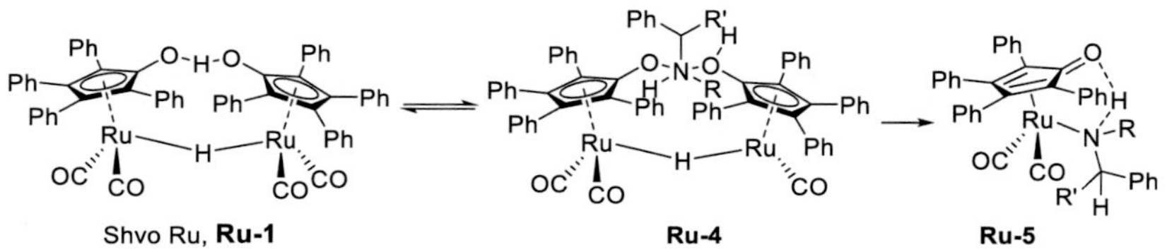

Scheme 36 Shvo catalyst activation by amines.

2.2.3 Ruthenium-catalyzed oxidation of diols and aminoalcohols

The authors envisioned that the biomimetic approach could be used for diol oxidation to access lactone derivatives under mild conditions. In 2011, Bäckvall's group disclosed a Ru-catalyzed aerobic lactonization of diols based on a biomimetic oxidation approach employing 2,6-dimethoxy-1,4-benzoquinone (DMBQ) and Co(salmdpt) complex as an efficient electron transfer mediator (Scheme 37) *(93)*. Notably, the lactonization process can be performed under air as well as under pure molecular oxygen to provide seven-, eight-, and nine-membered lactones in high yields without any side-reactions or byproducts. A variety of diols could be oxidized to give the corresponding lactones in high yields. Following this study, the Bäckvall's group extended the biomimetic approach to intramolecular oxidation for useful lactam synthesis from amino alcohols in 2012 *(94)*. The mechanism of these reactions proposed by the authors involves two oxidation steps; the first oxidation step involves of dehydrogenation of one alcohol **98** to

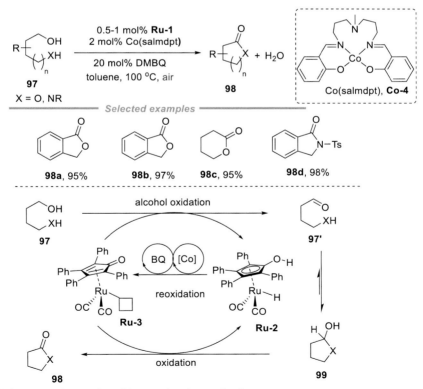

Scheme 37 Ru-catalyzed intramolecular cyclization.

generate aldehyde **97′** followed by intramolecular cyclization of other alcohol or amine with aldehyde to lactol or hemiaminal, which subsequently is oxidized to give the corresponding lactone or lactam. The active catalyst species **Ru-2** is regenerated after each oxidation step by reaction with benzoquinone derivative, cobalt complexes and molecular oxygen.

2.2.4 Ruthenium-catalyzed amine oxidation and cyclization

Later, the Bäckvall's group extended the biomimetic oxidations to coupling of benzylamines with 2-aminophenols, 2-aminobenzthiols, or 2-aminoanilines for the synthesis of benzoxazoles, benzothiazoles and benzimidazoles (Scheme 38) *(95)*. The method features a broad substrate scope and high functional group tolerance. This biomimetic method could be useful for the synthesis of pharmaceutically important heterocycles under mild reaction conditions. Mechanistically, the reaction was proposed to proceed through an imine intermediate. In the first step, dehydrogenation of

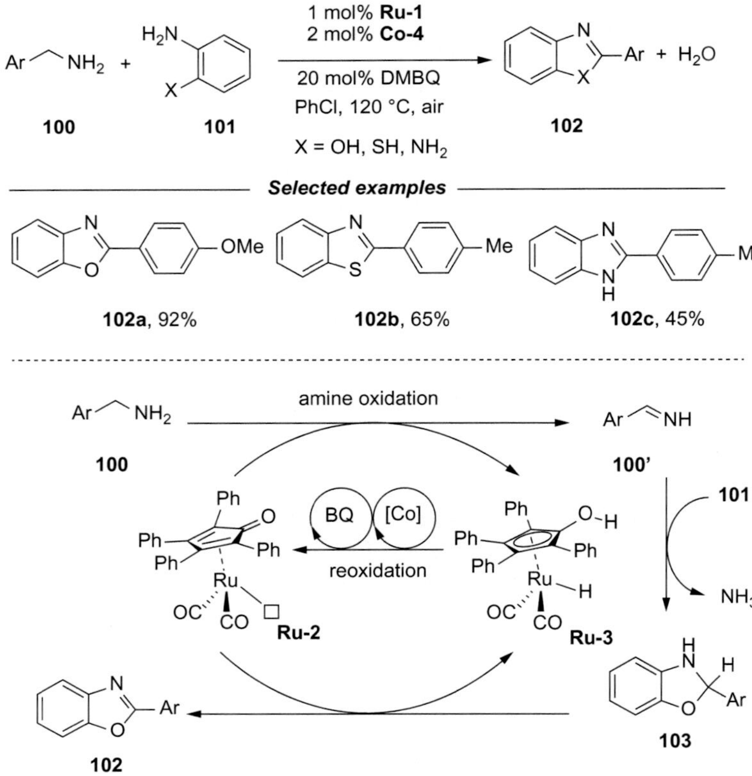

Scheme 38 Bäckvall's Ru-catalyzed coupling of benzylamines and 2-aminophenols.

the amine to give the corresponding imine followed by condensation of the imine with aminophenol with release of ammonia would generate 2-phenyldihydrobenzaoxazole intermediate **103**. Then the cyclic intermediate **103** leads to formation of benzoxazole after oxidation. After each oxidation step, the active catalyst species **Ru-2** is regenerated through the reaction of 2,6-dimethoxy-1,4-benzoquinone **94b** with Co(salmdpt) **Co-4** and molecular oxygen.

2.2.5 Ruthenium-catalyzed heterocycle oxidation

In the past few decades, mild and efficient dehydrogenations of alcohols and amines to give the corresponding aldehydes, ketones, imines, or heterocycles by the use of low-valent ruthenium complexes have been reported by Bäckvall's group *(19)*. Oxidation of N-heterocycles to their corresponding aromatic heterocycles is of great importance in natural products synthesis, materials science and medicinal chemistry. In 2014, the Stahl group developed a ruthenium-catalyzed aerobic dehydrogenation of diverse tetrahydroquinolines with ambient air to afford various aromatic heterocycles with good to excellent yields (Scheme 39) *(96)*. In this work, a ruthenium

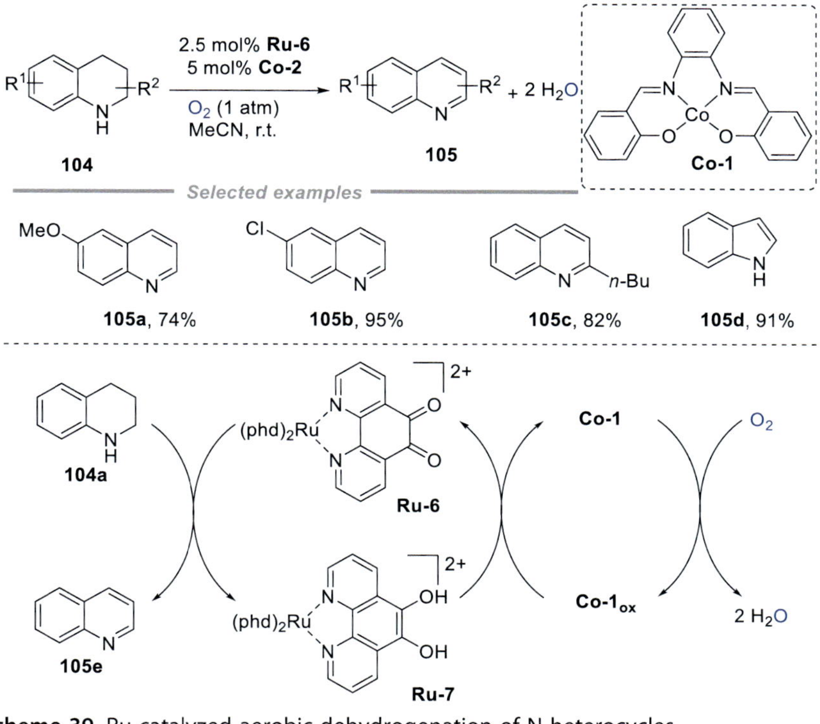

Scheme 39 Ru-catalyzed aerobic dehydrogenation of N-heterocycles.

complex $[Ru(phd)_3]^+$ (phd = 1,10-phenanthroline-5,6-dione) was successfully applied as an efficient catalyst and the structure of $[Ru(phd)_3]^+$ was characterized via X-ray crystallography. Notably, Co(salophen) **Co-1** was found to be an effective ETM in this process which allows the reaction to proceed efficiently. A wide variety of N-heterocycles were oxidized under the developed reaction conditions and the methodology was applied to various medicinally relevant quinolines. According to the proposed mechanism, first **Ru-6** dehydrogenates N-tetrahydroquinoline to provide quinoline and the dihydroxy ruthenium complex **Ru-7**. Next the complex **Ru-7** is oxidized by an oxidized form of Co(salophen) (**Co-1$_{ox}$**). The reduced form of Co(salophen) (**Co-1**) reacts with molecular oxygen to give an oxidized Co(salophen) (**Co-2$_{ox}$**) and H_2O.

2.3 Iron-catalyzed biomimetic oxidations with ETMs

Iron-catalyzed oxidations of organic substrates are of great importance in organic chemistry, because iron catalysts are inexpensive and environmentally friendly reagents. Over the past few decades numerous transition metal-catalyzed oxidation reactions have been developed (97). In spite of this development, there is an increasing demand for more selective, mild, and efficient oxidation methods using low-cost environmentally friendly catalysts. However, oxidation processes inspired by biological systems employing environmentally friendly and inexpensive oxidants such as molecular oxygen (O_2) are in high demand. In the following section different aspects of reaction methodology utilizing iron complexes will be covered while discussing the contributions that are categorized by the substrate type.

2.3.1 Iron-catalyzed alcohol oxidation

In 2020, the group of Bäckvall reported the first example of an iron-catalyzed biomimetic oxidation of alcohols using 2,6-dimethoxy-1,4-benzoquinone (DMBQ) and cobalt complex **Co-2** as electron transfer mediators (Scheme 40B) (98). The process is inspired by the electron transport chain (respiratory chain) used in living organisms and involves coupled redox processes that lead to a low-energy pathway through a stepwise oxidation using electron transfer mediators (ETMs) with O_2 as the terminal oxidant. An outer-sphere mechanism for dehydrogenation of the alcohol was proposed for the iron-catalyzed oxidations. The suggested mechanism proposed by the authors involves initial formation of the active catalyst **Fe-2a** species by reaction with TMANO with **Fe-1a**. The **Fe-2a** species then

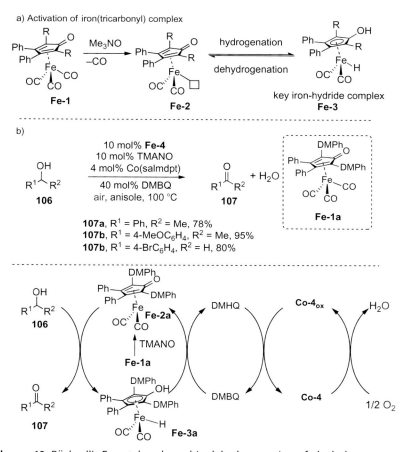

Scheme 40 Bäckvall's Fe-catalyzed aerobic dehydrogenation of alcohols.

reacts with the alcohol to produce the carbonyl compound and the iron hydride complex **Fe-3a** (**Scheme 40A**). The active iron catalyst **Fe-2a** is regenerated after reaction of the iron hydride intermediate **Fe-3a** with quinone DMBQ. The hydroquinone (2,6-dimethoxyhydroquinone, DMHQ) formed in this process is oxidized to DMBQ by an oxidized form of the cobalt macrocycle (**Co-4$_{ox}$**). The reduced form of cobalt macrocycle (**Co-4**) reacts with molecular oxygen to give an oxidized cobalt macrocycle (**Co-4$_{ox}$**) and H$_2$O. In this method, various primary and secondary alcohols were oxidized to their corresponding aldehydes and ketones with good to excellent yields.

2.3.2 Iron-catalyzed amine oxidation

Oxidation of amines to the corresponding imine compounds is an important class of reactions in organic chemistry. Amines are versatile and useful reagents in many organic transformations such as nucleophilic addition reaction, preparation of secondary and tertiary amines, and in cycloaddition reactions (99). From an environmental and cost-effective point of view, iron-catalyzed aerobic oxidation of amines to the corresponding imines are valuable and particularly attractive. In 2021, Bäckvall and co-workers developed a method for aerobic oxidation of amines to imines employing an iron catalyst and a bifunctional hybrid catalyst (Co(Salophen)-HQ) **Co-3** as an efficient electron transfer mediator (Scheme 41) (100). The electron transfer from the amine to molecular oxygen occurs using an iron

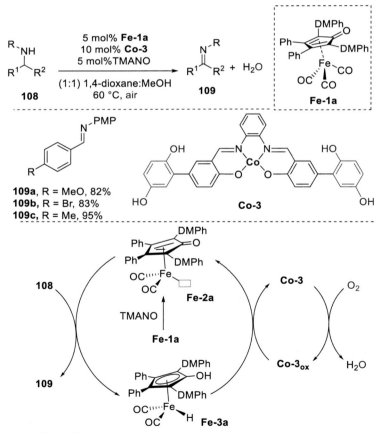

Scheme 41 Bäckvall's Fe-catalyzed aerobic dehydrogenation of amines.

complex and a bifunctional ETM coupled catalytic redox systems, leading to a low-energy pathway. A range of amines were oxidized to their corresponding imines in good to excellent yields. The combination of the Co-Schiff base and the quinone into a bifunctional hybrid catalyst provided an efficient electron transfer mediator for the iron-catalyzed oxidation of amines under aerobic conditions. Notably, it was observed that only trace amount of the desired product was formed when two separate ETMs (DMBQ and Co(salmdpt)) were used in the iron-catalyzed oxidation of amines. In this reaction, the solvent plays an important role. A 1:1 mixture of MeOH and 1,4-dioxane was found to be a useful solvent combination that provides an excellent yield (>95%, at 60 °C) with low catalyst loading. Pure 1,4-dioxane as solvent, however, provided a lower yield (40%, at 80 °C) from 4-methoxy-N-(4-methylbenzyl)aniline (**108c**) with high catalyst loading. In the mechanism suggested by the authors the active catalyst **Fe-2a** initially formed from **Fe-1a** reacts with **108** to furnish the iron hydride complex **Fe-3a** and imine **109**. The active iron catalyst **Fe-2a** is regenerated after reaction with iron hydride intermediate **Fe-3a** and an oxidized form of the hybrid catalyst (**Co-3$_{ox}$**). The reduced form of the hybrid catalyst (**Co-3**) reacts with molecular oxygen to give an oxidized hybrid catalyst (**Co-3$_{ox}$**) and H$_2$O.

2.3.3 Iron-catalyzed N-heterocycle oxidation

Catalytic oxidations of N-heterocycles to produce useful aromatic heterocycles constitute an importance and fundamental class of reactions in organic chemistry. Heterocycles presents many bioactive molecules, natural product and materials *(101)*. Oxidation of N-heterocycles to aromatic heterocycles inspired by biological systems employing an economical, environmentally friendly and inexpensive oxidant such as molecular oxygen (O$_2$) are increasing in demand. Aerobic oxidation using earth abundant and cheap iron-catalysts have become popular and are in high demand in organic synthesis *(102)*. Over the past few decades numerous transition metal-catalyzed oxidation reactions of N-heterocycles have been reported, however, there is an increasing demand for the improvement to obtain mild, efficient and scalable methods. In 2021, Bäckvall and co-workers disclosed an iron-catalyzed biomimetic aerobic of numerous N-heterocycles to afford aromatic heterocycles using a bifunctional hybrid catalyst (Co(Salophen)-HQ) **Co-3** as an efficient electron transfer mediator (Scheme 42) *(103)*. The method was found to allow catalytic oxidation of various N-heterocycles to their corresponding aromatic heterocycles in good to excellent yields. Notably, the authors found that the hybrid hydroquinone/cobalt Schiff base complex

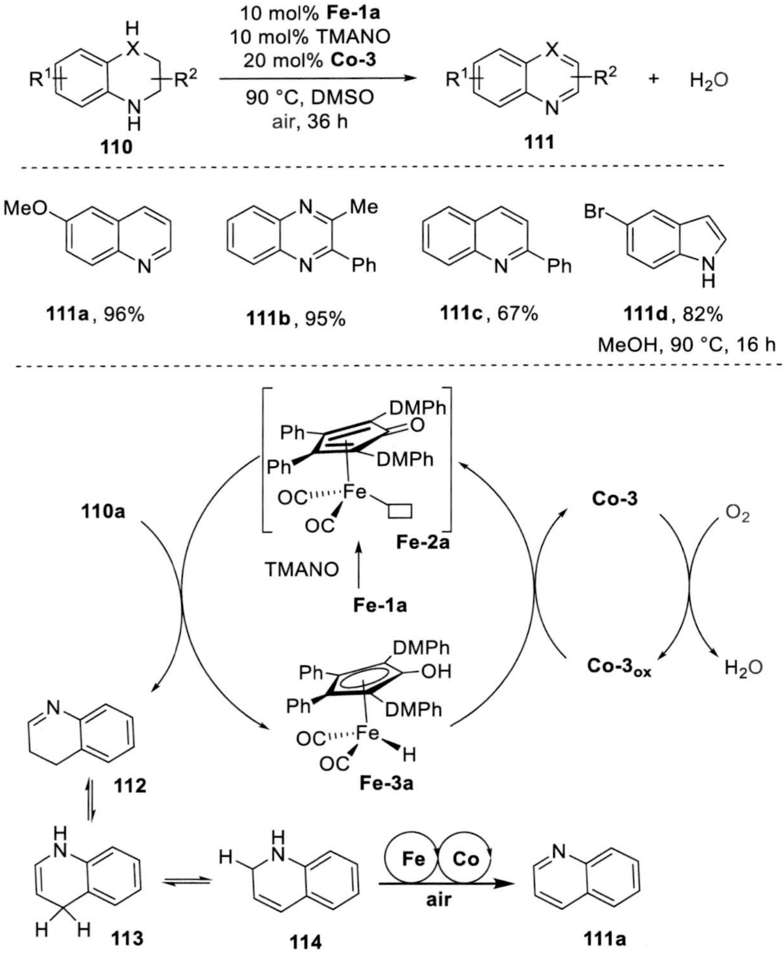

Scheme 42 Bäckvall's Fe-catalyzed aerobic oxidation of N-heterocycles.

provided the best result, with yield up to 94%, whereas the use of separate ETMs hydroquinone and Co(salophen) afforded the desired product in 75% yield. It was speculated that the intramolecular oxidation of hydroquinone to benzoquinone by oxidized Co(salophen) is much faster in the hybrid catalyst than intermolecular oxidation. The mechanism proposed by the authors for the oxidation of *N*-heterocycles is similar to that occurring in amine oxidation. According to the proposed mechanism, the active catalyst **Fe–2a** initially reacts with **110a** to furnish the iron hydride complex **Fe–3a** and imine intermediate **112**. The isomerization of imine **112** to intermediate

114 via **113** followed by aerobic oxidation catalyzed by the Fe—Co system afforded product **111a**. The active iron catalyst **Fe-2a** is regenerated after reaction with hydride intermediate **Fe-3a** and an oxidized form of the hybrid catalyst **Co-3**$_{ox}$. The reduced form of hybrid catalyst **Co-3** reacts with molecular oxygen to give an oxidized hybrid catalyst **Co-3**$_{ox}$ and H$_2$O.

3. Aerobic oxidation by biomimetic metal complexes
3.1 Iron-catalyzed biomimetic oxidation

Non-heme or heme iron complexes play important roles in aerobic oxidation such as epoxidations, C—H oxidations and many other reactions *(104)*. In this part, the central topic of our discussion will focus on recent reports of iron-catalyzed biomimetic oxidation of organic substrates employing molecular oxygen.

3.1.1 Iron-catalyzed epoxidation

Non-heme iron complexes have been utilized for epoxidation of alkenes in the past few decades. In 2011, the groups of Beller and Costas reported that an iron complex activated by molecular oxygen is capable of providing epoxide from variously substituted alkenes under air as the oxidant *(105)*. The reaction proceeds with fair to excellent chemoselectivity under mild conditions (Scheme 43). Mechanistic investigations suggest that epoxidation occurs via a radical-based autoxidation pathway. First, molecular oxygen is activated through co-substrate-assisted activation of O$_2$ to generate the active iron species **120**, which is responsible for the epoxidation of alkenes. This process represents an example of a highly challenging four-electron reduction of the O$_2$ molecule for the epoxidation of olefins. It is an example of a unique biomimetic catalyst system which is an attractive alternative for bioinspired catalysis. According to the proposed mechanism, the FeCl$_3$ first promotes the formation of active [Fe^{III}(**117**)$_2$**118**)]$^+$ complex **119**. Then the complex **119** is activated to generate the peroxide intermediate **120** that is responsible for the epoxidation of the olefin. Ethyl 2-oxocyclopentane-1-carboxylate **117** acts as co-substrate which is responsible for selective O$_2$ activation, generating the organic peroxide species **120**. In contrast, in the absence of alkene, the product could be formed via two oxygen atoms insertions into the β-keto ester **117** to furnish compound **122**, in a formal four-electron oxidation.

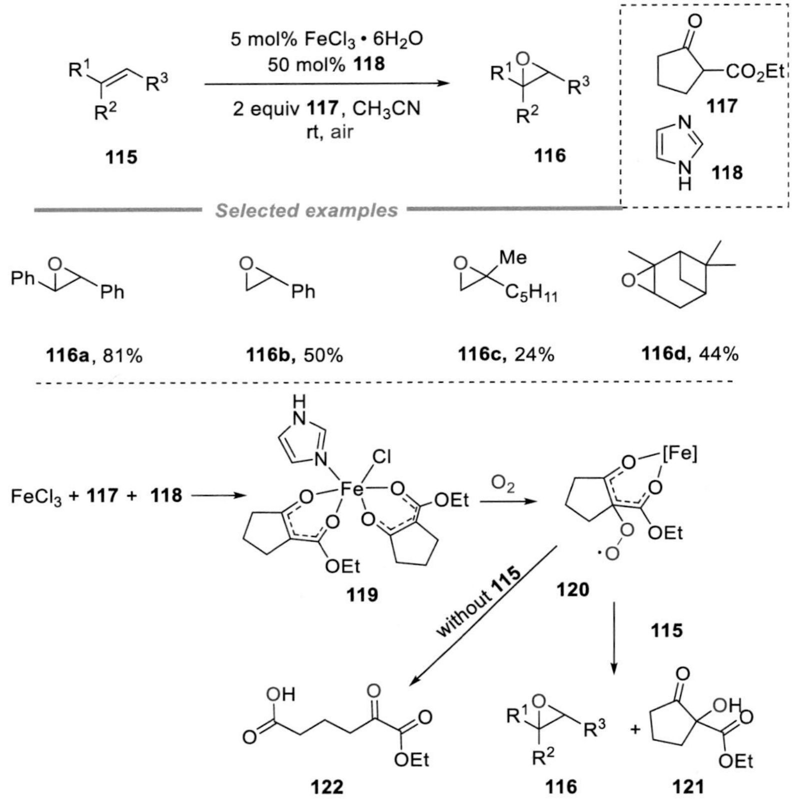

Scheme 43 Beller's Fe-catalyzed aerobic epoxidation of alkenes.

3.1.2 Iron-catalyzed C—H oxidation by non-heme iron

In 2014, Xiao and co-workers reported the Fe(II)-catalyzed selective α-oxidation of ethers under aerobic conditions (Scheme 44) *(106)*. In this method, tridentate pyridine bis(sulfonylimidazolidine) ligand-based iron catalysts were used for the chemoselective α-oxidation of a series of cyclic ethers by molecular oxygen. The methodology was used for the selective oxidation of cyclic ethers to afford valuable γ-butyrolactones, isochromanones, and phthalides with high chemoselectivity in good to excellent yields. It was found that the yields of the oxidation products strongly depend on the electronic effects of the substituents on the aromatic ring. Substrates containing electron-withdrawing substituents provided the highest catalytic turnover numbers, whereas the presence of electron-donating substituents leads to the opposite outcome. In the mechanism of the process proposed by the authors, the Fe(OTf)$_2$-**L** complex may first coordinate to one or two ethereal substrates to form intermediate **126** in which reaction with O$_2$ atmosphere would easily result in the formation of an Fe^{III} superoxo species **127**. In the following step, the Fe^{III} superoxo species leads to the cleavage of the α-CH bond of the Fe-bound ether substrate to furnish intermediate **128** via hydrogen atom transfer to the Fe^{III} center. Intermediate **128** provides a Fe^{IV}-(H)$_2$ dihydride and intermediate **129** through intramolecular hydrogen atom transfer from the α-hydrogen of the Fe-bound second ether molecule. The intermediate **129** is most probably formed by radical combination of α-radical intermediate of second ether and ether-peroxo radical. The dihydride Fe^{IV}-(H)$_2$ undergoes reductive elimination to give off H$_2$ and Fe^{II} complex. In the next step, the oxidative addition of the peroxide bond of **129** to Fe^{II} followed by β-hydrogen abstraction provides the ester product **124c** and H$_2$.

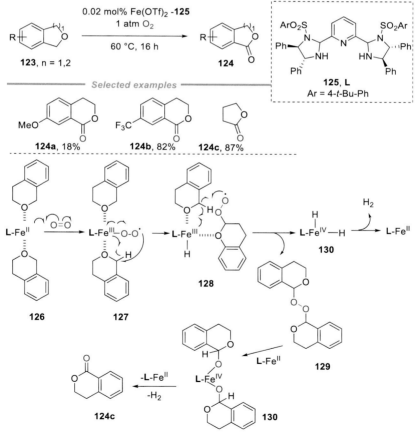

Scheme 44 Xiao's Fe-catalyzed aerobic oxidation of ethers.

3.1.3 Iron-catalyzed C—H oxidation by heme-complex

In 1994, Gray and co-workers developed the oxidation of 3-methylpentane to 3-hydroxy-3-methylpentane using iron-haloporphyrins and molecular oxygen as the sole oxidant (Scheme 45) *(107)*. The results of selective oxidation of **131** (>99%) and radical trapping experiments suggest that this reaction occurs via an autooxidation pathway. It was observed that oxidation by O_2 and Fe-(TFPPBr$_8$)Cl **133** provides higher selectivity (>99%) toward the tertiary position whereas the use of PhIO and Fe-(TFPPBr$_8$)Cl **133** affords lower selectivity. The mechanism proposed by the authors involves a reaction occurring through a metal-catalyzed radical-chain autooxidation. Initially, the alkane is oxidized to an alkyl radical and an alkyl hydroperoxide intermediate. The role of the metal is to catalyze the alkyl hydroperoxide decomposition to alkyl peroxide and alkoxy radicals. Radical trapping experiments led to inhibition by using 2,6-di-*tert*-butyl-4-methylphenol.

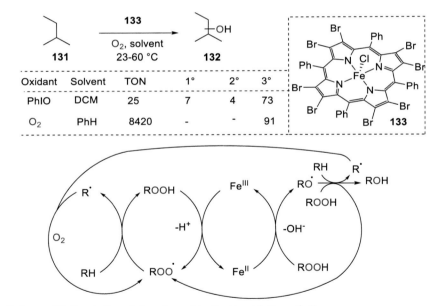

Scheme 45 Gray's Fe-catalyzed aerobic C—H oxidation of alkanes.

3.2 Manganese-catalyzed biomimetic oxidation
3.2.1 Manganese-catalyzed C—H oxidation

In 2018, the group of Zhou and Ji reported the biomimetic aerobic oxidation of toluene derivatives using a manganese porphyrins (TPPMnCl) as a catalyst (Scheme 46) *(108)*. Various toluene derivatives were oxidized.

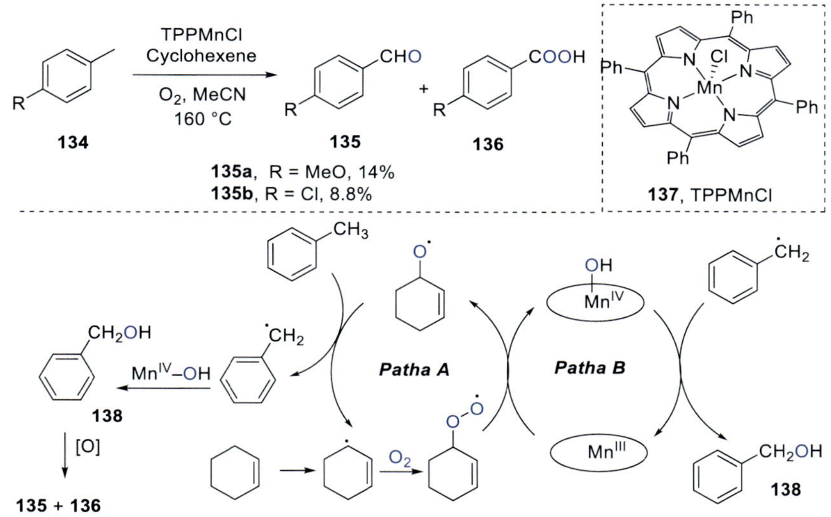

Scheme 46 Zhou and Ji's Mn-catalyzed aerobic C—H oxidation of toluenes.

Cyclohexene served as the key activator for dioxygen. Mechanistically, it was proposed by the authors that the reaction was initiated from the autoxidation of cyclohexene under heat to generate a cyclohexenyl radical intermediate, followed by the formation of cyclohexenyl peroxyl radical via trapping of cyclohexenyl radical with molecular oxygen. In the following step, Mn(III)–Cl **137** reacts with cyclohexenyl peroxyl radical to furnish active Mn(IV)–OH species and cyclohexenyl alkoxy radical, which reacts with toluene to produce a benzyl radical. Afterward, the benzyl radical reacts with high–valent active species (Mn(IV)–OH) to produce oxidized product. The corresponding oxidized products is formed in the presence of catalyst.

3.2.2 Manganese-catalyzed aerobic sulfoxidation

Manganese-catalyzed asymmetric aerobic sulfoxidations were demonstrated by the group of Mukaiyama in 1995 using molecular oxygen as the terminal oxidant (Scheme 47) *(109)*. A family of β-oxo–aldimino–manganese(III) complexes **141** allows asymmetric aerobic sulfoxidations in the presence of pivaldehydes **142** as sacrificial reductants. The reaction does not proceed in absence of aldehyde due to that the manganese(III) complexes is not activated by molecular oxygen directly. The developed methods allow to access various sulfoxides with moderate ee's and yields of products.

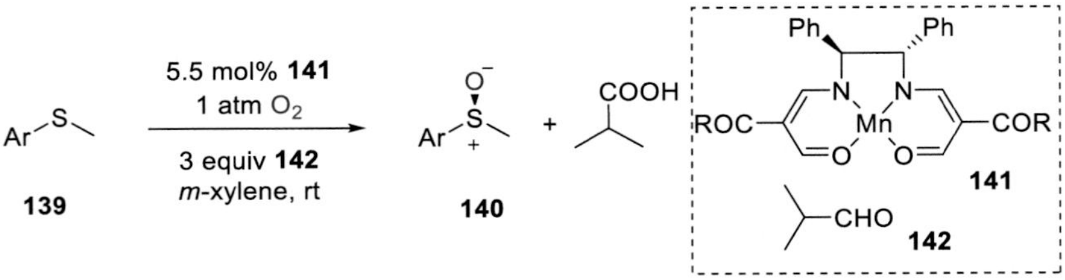

Scheme 47 reaction data:

substrate	catalyst	yield (%)	ee (%)
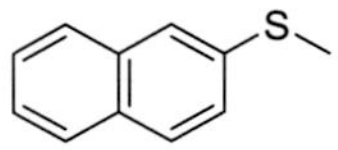 (2-naphthyl–S–)	R = O-cyclo-C$_5$H$_9$	57	51
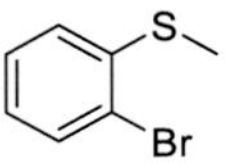 (2-bromophenyl–S–)	R = OEt	93	70
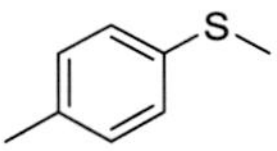 (4-methylphenyl–S–)	R = O-Me	58	44

Scheme 47 Mukaiyama Mn-catalyzed aerobic sulfoxidation.

3.3 Copper-catalyzed biomimetic oxidation

3.3.1 Copper-catalyzed alcohol oxidation

One of the earliest examples is a CuCl/TEMPO-catalyzed aerobic oxidation of primary benzylic and allylic alcohols reported by Semmelhack and co-workers in 1984 *(110)*. However, the oxidation of aliphatic alcohols required the use of stoichiometric CuCl$_2$. In 2003, Sheldon and co-workers developed similar type of reaction using a Cu^{II} salt with 2,2′-bipyridine (bpy) as a ligand and KOt-Bu as a base under ambient air as the oxidant *(111)*. Subsequently, Stahl and co-workers in 2011 developed a highly practical (bpy)Cu^I/TEMPO catalyst system for the selective oxidation of primary alcohols to aldehydes under air (Scheme 48A) *(112)*. A TEMPO-free copper-catalyzed aerobic oxidation of primary and secondary alcohols was developed by the group of Lumb and Arndtsen in 2015 employing Cu/diamine complex inspired by the enzyme tyrosinase *(113)*. In 2017, Lumb, Arndtsen, and Stahl and co-workers disclosed a nitroxyl-free copper-catalyzed aerobic alcohol oxidation to afford carbonyl under oxygen atmosphere (Scheme 48B) *(114)*. It was proposed that the oxygenation of ligand **146** in the oxidative self-processing step generates N-hydroxylamine intermediate **148**. In the following step, intermediate **148** is oxidized by

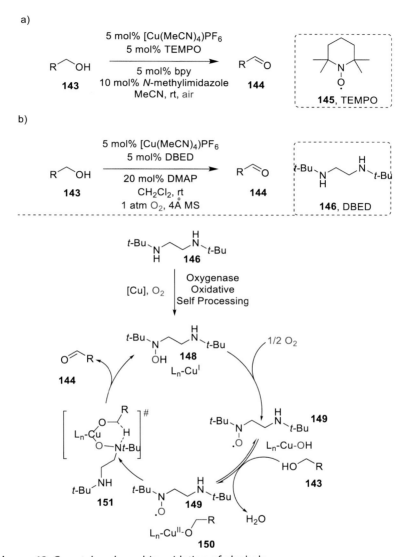

Scheme 48 Cu-catalyzed aerobic oxidation of alcohols.

molecular oxygen and L_n-Cu^I to furnish nitroxyl radical **149** and a L_n-Cu-OH intermediate. Alcohol **143** reacts with a L_n-Cu-OH intermediate to generate copper-alkoxy complex **150** via ligand exchange. The nitroxyl radical is responsible for the oxidation of alcohol via transition-state **151**. The authors also suggested Cu/O$_2$ adduct is responsible for the in-situ oxygenation of DBED to generate **148**. The oxygenated DBED serves as a co-catalyst in an oxidase-type catalytic cycle.

3.3.2 Copper-catalyzed phenol oxidation

In 2014, Lumb and co-workers reported a practical catalytic aerobic oxygenation of phenols to *ortho*-quinones *(115)*. It was found by the authors that catalytic quantities of $[Cu(CH_3CN)_4]PF_6$ in conjunction with *N,N*-di-*tert*-butylethylenediamine **146** and O_2 was able to promote oxidation of the *ortho*-position of phenols. The proposed mechanism is shown in Scheme 49. The characteristic $\mu\text{-}\eta^2,\eta^2$ peroxo di-copper reactive oxygen species **146** is generated in the reductive activation of by dioxygen in which a mimic of the type II Cu enzyme tyrosinase. This active species can *ortho*-oxygenate the phenol into **L**Cu(II)-semiquinone, followed oxidative C—O coupling to provide *ortho*-quinones. Complementary substrate scope and functional group tolerance to the Cu(I)-catalyzed aerobic functionalization of phenols was noted.

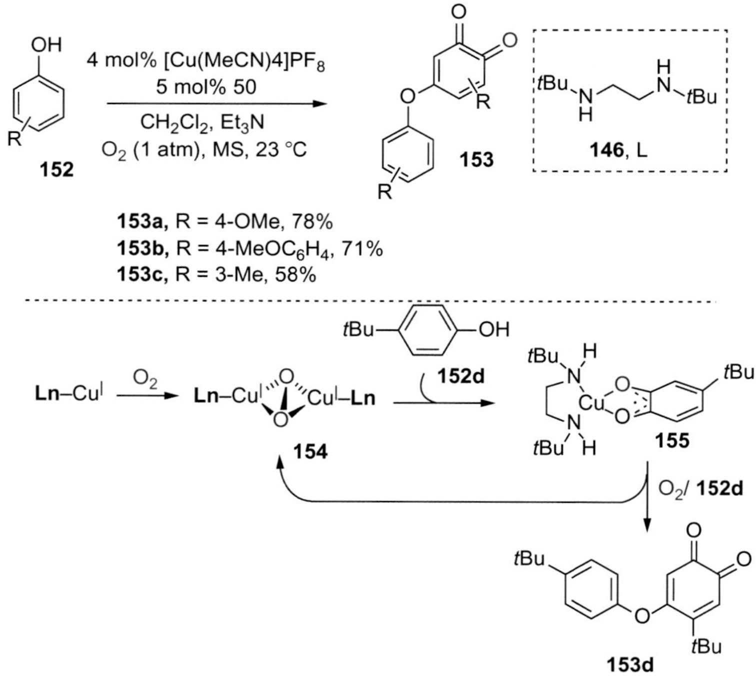

153a, R = 4-OMe, 78%
153b, R = 4-MeOC$_6$H$_4$, 71%
153c, R = 3-Me, 58%

Scheme 49 Lumb's Cu-catalyzed aerobic oxidation of phenols.

3.4 Osmium-catalyzed biomimetic oxidation

In 2020, Wang, Xia, and coworkers reported a stable osmium–peroxo-catalyzed aerobic oxidation of alcohol with molecular oxygen affording the corresponding aldehyde **157** (Scheme 50) *(116)*. Firstly, the osmium-peroxo species (**160**) reacts with alcohol to furnish aldehyde and osmium-hydrido-hydroperoxo via transition state (**TS–161**). Subsequently, the second active osmium-oxo-complex **164** can be generated from intermediate **163** via heterolytic cleavage of the O—O bond. Intermediate **164** reacts with alcohol to furnish aldehyde and catalytic intermediate **165**. Transition state (**TS–161**) is regenerated in the presence of molecular oxygen and alcohol. The authors suggested that the turn-over limiting step involves the transfer of a hydrogen atom from the benzylic CH_2 of the alcohol which is supported by the kinetic result and theoretical calculation.

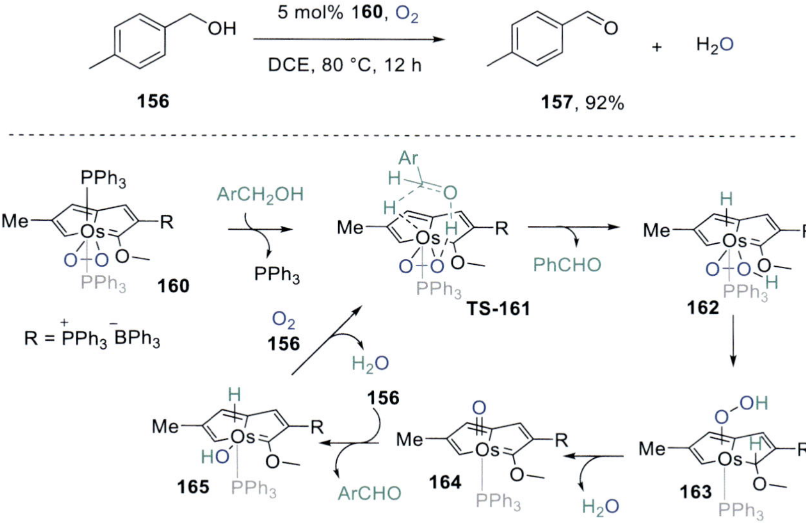

Scheme 50 Wang and Xia's Os-catalyzed aerobic oxidation of alcohols.

3.5 Vanadium-catalyzed biomimetic oxidation

In 2014, Ogawa and co-workers developed a vanadium (IV)–catalyzed aerobic green oxidation of alcohol using water as the sole solvent under atmospheric pressure of O_2 (Scheme 51) *(117)*. This operationally simple method was applied to gram-scale synthesis under an air atmosphere and this method was found to be useful for green oxidation in organic synthesis and

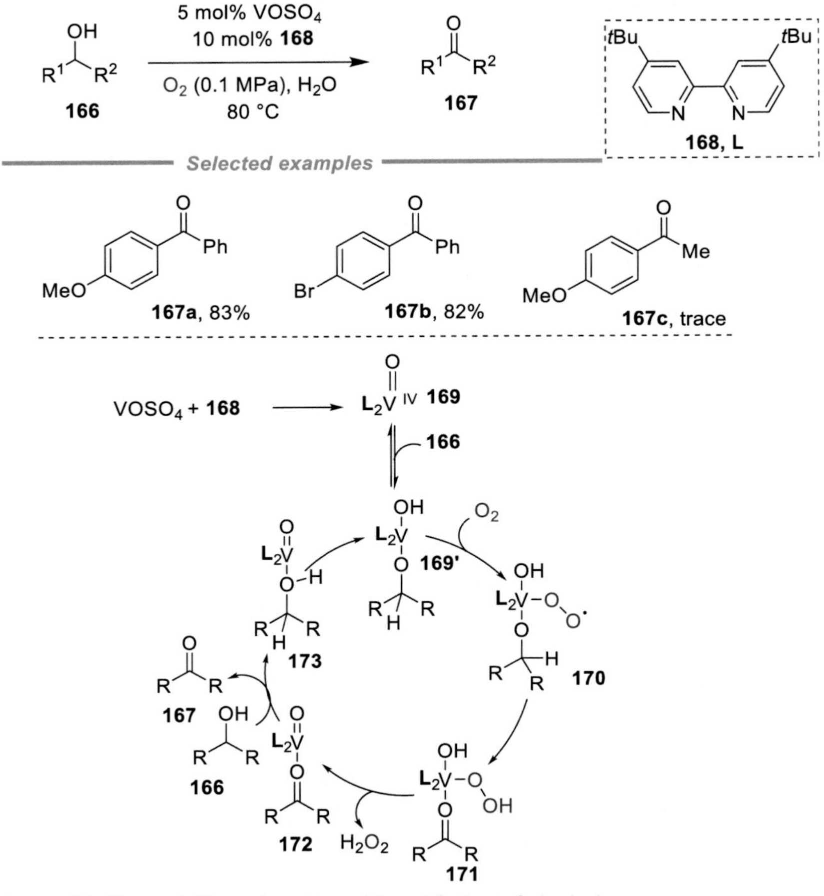

Scheme 51 Ogawa's V-catalyzed aerobic oxidation of alcohols.

low-costly industrial chemical applications. Electron-rich and electron-neutral aryl/alkyl-substituted secondary alcohols were oxidized to their corresponding carbonyl compounds. The reaction mechanism proposed by the authors is shown in Scheme 51. First, the complex **169′** is generated by the reaction of alcohol with preformed vanadium complex **169**. Complex **169′** next generates vanadium peroxide complex **170** by oxidation with molecular oxygen. Afterward, the complex **171** is formed through intramolecular hydrogen abstraction by vanadium peroxide radical (**170**). Release of H_2O_2 from intermediate **171** generates complex **172**. In the following step, the ligand exchange by alcohol provides ketone, which completes the catalytic cycle. ^{51}V NMR analysis revealed the formation of vanadium(V) species in this reaction.

In 2012, the Hanson and Scott group showcased a homogeneous vanadium(V)-catalyzed aerobic oxidation of alcohol to afford aldehyde (Scheme 52) *(118)*. The computational and experimental studies suggested that the alcohol oxidation was catalyzed by vanadium complex **176** in the presence of external base. Most probably the vanadium alcoholate complexes **177** is formed via ligand exchange with **176** and benzylic alcohol. The base plays a key role in the reaction, which was supported by control experiment. Only 12% conversion was observed in absence of NEt$_3$ as a base. In the presence of NEt$_3$, benzaldehyde (0.5 equivalents), and benzyl alcohol (0.5 equivalents) complex **177** was generated quantitatively [(HQ) 2V^{IV}(O)]. Notably, it was observed that the reaction is first order in both **174** and NEt$_3$ with activation parameters $\triangle H^{\neq}$ (28±4) kJ^{-1} mol^{-1} and $\triangle S^{\neq}$ (−169±4) JK^{-1} mol^{-1}.

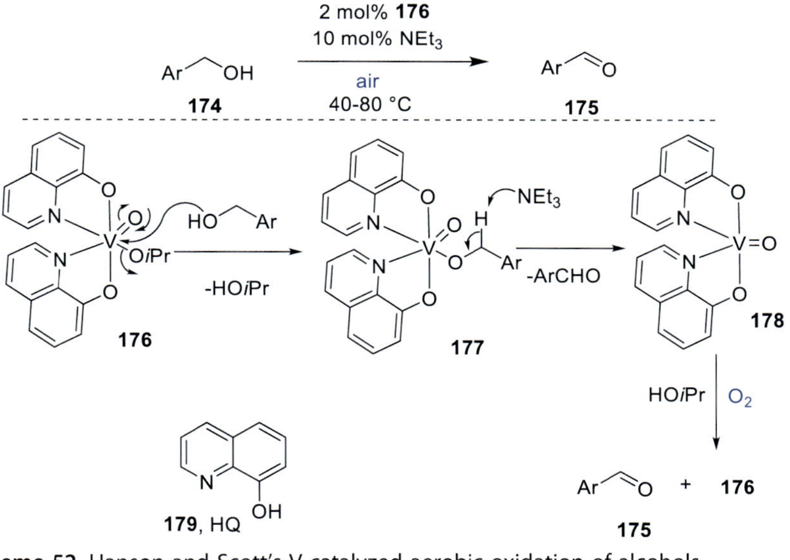

Scheme 52 Hanson and Scott's V-catalyzed aerobic oxidation of alcohols.

4. Concluding remarks and future perspectives

The current chapter attempts to provide a thorough review of biomimetic oxidation of organic substrates using various transition metal catalysts under aerobic conditions. We have covered a variety of biomimetic oxidations of organic substrates carried out by our group and other notable researchers in the field. Transition metal-catalyzed aerobic biomimetic oxidation has made it possible to create synthetically valuable products from easily available materials in recent years. The development of transition metal-catalyzed aerobic oxidation mimicking reactions occurring in Nature has received significant attention because of their environmentally friendly character. Although many elegant methods have been developed in this field, several challenges have yet to be overcome. It is desirable that most of the aerobic biomimetic oxidations are based on the use of first row transition metals. Furthermore, the barriers for electron transfer between the substrate to be oxidized and molecular oxygen has to become even lower.

We believe that biomimetic aerobic oxidative transformations have a lot of intriguing prospects and possibilities in the future. More diverse green and mild biomimetic methods for the oxidation of various organic substrates will bridge the gap in this chemical space. Future studies should see further development of biomimetic oxidations employing solid-supported metal complexes to improve the efficiency of processes and simplify product isolation. Aerobic biomimetic oxidations that are both regio- and chemoselective would surely be a potent platform for the synthesis of complex molecular structures. Biomimetic approaches for late-stage oxidation remain unsolved puzzles. In the near future, aerobic biomimetic enantioselective oxidation will be accomplished and applied to chemical industry and natural product synthesis.

Acknowledgments

Financial support from the Swedish Research Council (2019-04042), the Olle Engkvist Foundation, and the Knut and Alice Wallenberg Foundation (KAW 2016.0072) is gratefully acknowledged.

References

1. Bäckvall, J.-E. *Modern Oxidation Methods*; John Wiley & Sons, 2011.
2. Jiao, N.; Stahl, S. S. *Green Oxidation in Organic Synthesis*; John Wiley & Sons, 2019.
3. Wood, P. M. *Biochem. J.* **1988**, *253*, 287.
4. Guo, M.; Corona, T.; Ray, K.; Nam, W. *ACS Cent. Sci.* **2019**, *5*, 13–28.
5. Jones, R. D.; Summerville, D. A.; Basolo, F. *Chem. Rev.* **1979**, *79*, 139–179.
6. Oloo, W. N.; Que, L. *Acc. Chem. Res.* **2015**, *48*, 2612–2621.
7. Martin, D. R.; Matyushov, D. V. *Sci. Rep.* **2017**, *7*, 1–11.
8. Moran, L. A.; Horton, R. A.; Scrimgeour, K. G.; Perry, M. D. *Principles of Biochemistry*; Pearson: London, 2014.
9. Cox, M. M.; Nelson, D. L. *Lehninger Principles of Biochemistry*, vol. 5; WH Freeman: New York, 2008.
10. Bäckvall, J.-E.; Awasthi, A.; Renko, Z. *J. Am. Chem. Soc.* **1987**, *109*, 4750–4752.
11. Liang, Y.; Wei, J.; Qiu, X.; Jiao, N. *Chem. Rev.* **2018**, *118*, 4912–4945.

12. Zografos, A.; Petsi, M. *Synthesis* **2018**, *50*, 4715–4745.
13. Huang, X.; Groves, J. T. *Chem. Rev.* **2018**, *118*, 2491–2553.
14. Vicens, L.; Olivo, G.; Costas, M. *ACS Catal.* **2020**, *10*, 8611–8631.
15. Que, L.; Tolman, W. B. *Nature* **2008**, *455*, 333–340.
16. Jiang, G.; Liu, Q.; Guo, C. In *Biomimetic Based Applications*; George, A., Ed.; IntechOpen, 2011.
17. Wang, D.; Weinstein, A. B.; White, P. B.; Stahl, S. S. *Chem. Rev.* **2017**, *118*, 2636–2679.
18. Piera, J.; Bäckvall, J.-E. *Angew. Chem. Int. Ed.* **2008**, *47*, 3506–3523.
19. Liu, J.; Guðmundsson, A.; Bäckvall, J.-E. *Angew. Chem. Int. Ed.* **2020**, *60*, 15686–15704.
20. Jira, R. *Angew. Chem. Int. Ed.* **2009**, *48*, 9034–9037.
21. Phillips, F. *Am. Chem. J.* **1894**, *16*, 255.
22. Smidt, J.; Hafner, W.; Jira, R.; Sedlmeier, J.; Sieber, R.; Rüttinger, R.; Kojer, H. *Angew. Chem.* **1959**, *71*, 176–182.
23. Eckert, M.; Fleischmann, G.; Jira, R.; Bolt, H. M.; Golka, K. In Ullmann's Encyclopedia of Industrial Chemistry, vol. 1; Wiley-VCH Verlag: Weinheim, Germany, 2000; pp. 191–207.
24. Kočovský, P.; Bäckvall, J.-E. *Chem. Eur. J.* **2015**, *21*, 36–56.
25. Tsuji, J.; Nagashima, H.; Nemoto, H. *Org. Synth.* **1984**, *62*, 9.
26. Tsuji, J. *Synthesis* **1984**, 369–384.
27. Clement, W. H.; Selwitz, C. M. *J. Organomet. Chem.* **1964**, *29*, 241–243.
28. Bäckvall, J.-E.; Hopkins, R. B. *Tetrahedron Lett.* **1988**, *29*, 2885–2888.
29. Bäckvall, J.-E.; Hopkins, R. B.; Grennberg, H.; Mader, M.; Awasthi, A. K. *J. Am. Chem. Soc.* **1990**, *112*, 5160–5166.
30. Grate, J. H.; Hamm, D. R.; Mahajan, S. *Polyoxometalates: From Platonic Solids to Antiretroviral Activity*; Kluwer: Dordrecht, 1994.
31. Dong, J. J.; Browne, W. R.; Feringa, B. L. *Angew. Chem. Int. Ed.* **2015**, *54*, 734–744.
32. Hosokawa, T.; Ohta, T.; Kanayama, S.; Murahashi, S. *J. Organomet. Chem.* **1987**, *52*, 1758–1764.
33. Weiner, B.; Baeza, A.; Jerphagnon, T.; Feringa, B. L. *J. Am. Chem. Soc.* **2009**, *131*, 9473–9474.
34. Choi, P. J.; Sperry, J.; Brimble, M. A. *J. Organomet. Chem.* **2010**, *75*, 7388–7392.
35. Feringa, B. L. *J. Chem. Soc. Chem. Commun.* **1986**, 909–910.
36. Wickens, Z. K.; Morandi, B.; Grubbs, R. H. *Angew. Chem. Int. Ed.* **2013**, *52*, 11257–11260.
37. Wickens, Z. K.; Skakuj, K.; Morandi, B.; Grubbs, R. H. *J. Am. Chem. Soc.* **2014**, *136*, 890–893.
38. Timokhin, V. I.; Stahl, S. S. *J. Am. Chem. Soc.* **2005**, *127*, 17888–17893.
39. Mitsudome, T.; Mizumoto, K.; Mizugaki, T.; Jitsukawa, K.; Kaneda, K. *Angew. Chem. Int. Ed.* **2010**, *49*, 1238–1240.
40. Morandi, B.; Wickens, Z. K.; Grubbs, R. H. *Angew. Chem. Int. Ed.* **2013**, *52*, 2944–2948.
41. Grennberg, H.; Faizon, S.; Bäckvall, J.-E. *Angew. Chem. Int. Ed.* **1993**, *32*, 263–264.
42. Ning, X.-S.; Wang, M.-M.; Qu, J.-P.; Kang, Y.-B. *J. Organomet. Chem.* **2018**, *83*, 13523–13529.
43. Ning, X.-S.; Liang, X.; Hu, K.-F.; Yao, C.-Z.; Qu, J.-P.; Kang, Y.-B. *Adv. Synth. Catal.* **2018**, *360*, 1590–1594.
44. Ren, W.; Xia, Y.; Ji, S.-J.; Zhang, Y.; Wan, X.; Zhao, J. *Org. Lett.* **2009**, *11*, 1841–1844.
45. Zhou, P.; Jiang, H.; Huang, L.; Li, X. *Chem. Commun.* **2011**, *47*, 1003–1005.
46. Zhou, P.; Zheng, M.; Jiang, H.; Li, X.; Qi, C. *J. Organomet. Chem.* **2011**, *76*, 4759–4763.
47. Wang, P.-S.; Gong, L.-Z. *Acc. Chem. Res.* **2020**, *53*, 2841–2854.

48. Pàmies, O.; Margalef, J.; Cañellas, S.; James, J.; Judge, E.; Guiry, P. J.; Moberg, C.; Backvall, J.-E.; Pfaltz, A.; Pericàs, M. A. *Chem. Rev.* **2021**, *121*, 4373–4505.
49. Henderson, W. H.; Check, C. T.; Proust, N.; Stambuli, J. P. *Org. Lett.* **2010**, *12*, 824–827.
50. Kozack, C. V.; Sowin, J. A.; Jaworski, J. N.; Iosub, A. V.; Stahl, S. S. *ChemSusChem* **2019**, *12*, 3003–3007.
51. Pattillo, C. C.; Strambeanu, I. I.; Calleja, P.; Vermeulen, N. A.; Mizuno, T.; White, M. C. *J. Am. Chem. Soc.* **2016**, *138*, 1265–1272.
52. Moritanl, I.; Fujiwara, Y. *Tetrahedron Lett.* **1967**, *8*, 1119–1122.
53. Fujiwara, Y.; Moritani, I.; Matsuda, M.; Teranishi, S. *Tetrahedron Lett.* **1968**, *9*, 633–636.
54. Fujiwara, Y.; Moritani, I.; Danno, S.; Asano, R.; Teranishi, S. *J. Am. Chem. Soc.* **1969**, *91*, 7166–7169.
55. Dwight, T. A.; Rue, N. R.; Charyk, D.; Josselyn, R.; DeBoef, B. *Org. Lett.* **2007**, *9*, 3137–3139.
56. Gigant, N.; Bäckvall, J.-E. *Org. Lett.* **2014**, *16*, 1664–1667.
57. Gigant, N.; Bäckvall, J.-E. *Org. Lett.* **2014**, *16*, 4432–4435.
58. Zultanski, S. L.; Stahl, S. S. *J. Organomet. Chem.* **2015**, *793*, 263–268.
59. Gao, Y.; Gao, Y.; Wu, W.; Jiang, H.; Yang, X.; Liu, W.; Li, C. *J. Chem. Eur. J.* **2017**, *23*, 793–797.
60. Ye, X.; Shi, X. *Org. Lett.* **2014**, *16*, 4448–4451.
61. Lu, M.-Z.; Chen, X.-R.; Xu, H.; Dai, H.-X.; Yu, J.-Q. *Chem. Sci.* **2018**, *9*, 1311–1316.
62. Zhang, J.; Lu, X.; Shen, C.; Xu, L.; Ding, L.; Zhong, G. *Chem. Soc. Rev.* **2021**, *50*, 3263–3314.
63. Liu, M.; Yang, P.; Karunananda, M. K.; Wang, Y.; Liu, P.; Engle, K. M. *J. Am. Chem. Soc.* **2018**, *140*, 5805–5813.
64. Stowers, K. J.; Kubota, A.; Sanford, M. S. *Chem. Sci.* **2012**, *3*, 3192–3195.
65. Ma, S. *Chem. Rev.* **2005**, *105* (7), 2829–2872.
66. Krause, N.; Hashmi, A. S. *Modern Allene Chemistry*; Wiley-VCH, 2004.
67. Piera, J.; Närhi, K.; Bäckvall, J.-E. *Angew. Chem. Int. Ed.* **2006**, *118*, 7068–7071.
68. Piera, J.; Persson, A.; Caldentey, X.; Bäckvall, J.-E. *J. Am. Chem. Soc.* **2007**, *129*, 14120–14121.
69. Volla, C. M.; Bäckvall, J.-E. *Angew. Chem. Int. Ed.* **2013**, *52*, 14209–14213.
70. Qiu, Y.; Yang, B.; Zhu, C.; Bäckvall, J.-E. *Chem. Sci.* **2017**, *8*, 616–620.
71. Zhu, C.; Yang, B.; Jiang, T.; Bäckvall, J.-E. *Angew. Chem. Int. Ed.* **2015**, *54*, 9066–9069.
72. Purse, B. W.; Tran, L. H.; Piera, J.; Åkermark, B.; Bäckvall, J.-E. *Chem. Eur. J.* **2008**, *14*, 7500–7503.
73. Johnston, E. V.; Karlsson, E. A.; Lindberg, S. A.; Åkermark, B.; Bäckvall, J.-E. *Chem. Eur. J.* **2009**, *15*, 6799–6801.
74. Persson, A. K.; Bäckvall, J.-E. *Angew. Chem. Int. Ed.* **2010**, *49*, 4624–4627.
75. Liu, J.; Ricke, A.; Yang, B.; Bäckvall, J.-E. *Angew. Chem. Int. Ed.* **2018**, *57*, 16842–16846.
76. Zhu, C.; Liu, J.; Mai, B. K.; Himo, F.; Bäckvall, J.-E. *J. Am. Chem. Soc.* **2020**, *142*, 5751–5759.
77. Liu, J.; Bäckvall, J.-E. *Chem. Eur. J.* **2020**, *26*, 15513–15518.
78. Trzeciak, A.; Augustyniak, A. *Coord. Chem. Rev.* **2019**, *384*, 1–20.
79. Mpungose, P. P.; Vundla, Z. P.; Maguire, G. E.; Friedrich, H. B. *Molecules* **2018**, *23*, 1676.
80. Biffis, A.; Centomo, P.; Del Zotto, A.; Zecca, M. *Chem. Rev.* **2018**, *118*, 2249–2295.
81. Li, M.-B.; Bäckvall, J.-E. *Acc. Chem. Res.* **2021**, *54*, 2275–2286.

82. Li, J.; Liao, J.; Ren, Y.; Liu, C.; Yue, C.; Lu, J.; Jiang, H. *Angew. Chem. Int. Ed.* **2019**, *58*, 17148–17152.
83. Kong, W.-J.; Reil, M.; Feng, L.; Li, M.-B.; Bäckvall, J.-E. *CCS Chem.* **2021**, 1127–1137.
84. Arends, I. W. C. E.; Sheldon, R. A. In *Modern Oxidation Methods*; Bäckvall, J.-E., Ed.; Wiley-VCH: Weinheim, 2004; p. 83.
85. Dijksman, A.; Marino-González, A.; Mairata i Payeras, A.; Arends, I. W. C. E.; Sheldon, R. A. *J. Am. Chem. Soc.* **2001**, *123*, 6826–6833.
86. Blum, Y.; Shvo, Y. *Isr. J. Chem.* **1984**, *24*, 144–148.
87. Wang, G.-Z.; Andreasson, U.; Bäckvall, J.-E. *J. Chem. Soc. Chem. Commun.* **1994**, 1037–1038.
88. Samec, J. S. M.; Éll, A. H.; Bäckvall, J.-E. *Chem. Eur. J.* **2005**, *11*, 2327–2334.
89. Mase, N. In *Comprehensive Chirality*; Carreira, E. M., Yamamoto, H., Eds.; Elsevier: Amsterdam, 2012; pp. 12–97.
90. Csjernyik, G.; Éll, A. H.; Fadini, L.; Pugin, B.; Bäckvall, J. E. *J. Organomet. Chem.* **2002**, *67*, 1657–1662.
91. Samec, J. S. M.; Éll, A. H.; Åberg, J. B.; Privalov, T.; Eriksson, L.; Bäckvall, J.-E. *J. Am. Chem. Soc.* **2006**, *128*, 14293–14305.
92. Warner, M. C.; Casey, C.-P.; Bäckvall, J.-E. *Top. Organomet. Chem.* **2011**, *37*, 85–125.
93. Endo, Y.; Bäckvall, J.-E. *Chem. Eur. J.* **2011**, *17*, 12596–12601.
94. Babu, B. P.; Endo, Y.; Bäckvall, J.-E. *Chem. Eur. J.* **2012**, *18*, 11524–11527.
95. Endo, Y.; Bäckvall, J.-E. *Chem. Eur. J.* **2012**, *18*, 13609–13613.
96. Wendlandt, A. E.; Stahl, S. S. *J. Am. Chem. Soc.* **2014**, *136*, 11910–11913.
97. Punniyamurthy, T.; Velusamy, S.; Iqbal, J. *Chem. Rev.* **2005**, *105*, 2329–2364.
98. Guðmundsson, A.; Schlipköter, K. E.; Bäckvall, J.-E. *Angew. Chem. Int. Ed.* **2020**, *59*, 5403–5406.
99. Martin, S. F. *Pure Appl. Chem.* **2009**, *81*, 195–204.
100. Guðmundsson, A.; Manna, S.; Bäckvall, J.-E. *Angew. Chem. Int. Ed.* **2021**, *60*, 11819–11823.
101. Matada, B. S.; Pattanashettar, R.; Yernale, N. G. *Biorg. Med. Chem.* **2021**, *32*, 115973.
102. Chen, X.; Chen, T.; Ji, F.; Zhou, Y.; Yin, S.-F. *Cat. Sci. Technol.* **2015**, *5*, 2197–2202.
103. Manna, S.; Kong, W.-J.; Bäckvall, J.-E. *Chem. Eur.* **2021**, https://doi.org/10.1002/chem.202102483.
104. Bauer, E. B. *Isr. J. Chem.* **2017**, *57*, 1131–1150.
105. Schröder, K.; Join, B.; Amali, A. J.; Junge, K.; Ribas, X.; Costas, M.; Beller, M. *Angew. Chem. Int. Ed.* **2011**, *50*, 1425–1429.
106. Gonzalez-de-Castro, A.; Robertson, C. M.; Xiao, J. *J. Am. Chem. Soc.* **2014**, *136*, 8350–8360.
107. Grinstaff, M.; Hill, M.; Labinger, J.; Gray, H. *Science* **1994**, *264*, 1311–1313.
108. Chen, H.-Y.; Lv, M.; Zhou, X.-T.; Wang, J.-X.; Han, Q.; Ji, H.-B. *Catal. Commun.* **2018**, *109*, 76–79.
109. Kiyomi, I.; Takushi, N.; Tohru, Y.; Teruaki, M. *Chem. Lett.* **1995**, *24*, 335–336.
110. Semmelhack, M. F.; Schmid, C. R.; Cortes, D. A.; Chou, C. S. *J. Am. Chem. Soc.* **1984**, *106*, 3374–3376.
111. Gamez, P.; Arends, I. W. C. E.; Reedijk, J.; Sheldon, R. A. *Chem. Commun.* **2003**, 2414–2415.
112. Hoover, J. M.; Stahl, S. S. *J. Am. Chem. Soc.* **2011**, *133*, 16901–16910.
113. Xu, B.; Lumb, J.-P.; Arndtsen, B. A. *Angew. Chem. Int. Ed.* **2015**, *54*, 4208–4211.
114. McCann, S. D.; Lumb, J.-P.; Arndtsen, B. A.; Stahl, S. S. *ACS Cent. Sci.* **2017**, *3*, 314–321.
115. Esguerra, K. V. N.; Fall, Y.; Lumb, J.-P. *Angew. Chem. Int. Ed.* **2014**, *53*, 5877–5881.

116. Deng, Z.; Wu, P.; Cai, Y.; Sui, Y.; Chen, Z.; Zhang, H.; Wang, B.; Xia, H. *iScience* **2020**, *23*, 101379.
117. Marui, K.; Higashiura, Y.; Kodama, S.; Hashidate, S.; Nomoto, A.; Yano, S.; Ueshima, M.; Ogawa, A. *Tetrahedron* **2014**, *70*, 2431–2438.
118. Wigington, B. N.; Drummond, M. L.; Cundari, T. R.; Thorn, D. L.; Hanson, S. K.; Scott, S. L. *Chem. Eur. J.* **2012**, *18*, 14981–14988.

About the authors

Srimanta Manna received his MSc from IIT Bombay (India) in 2012, and his PhD from the Max Planck Institute of Molecular Physiology, Dortmund (Germany) in 2017. His doctoral studies were carried out in the group of Prof. Andrey P. Antonchick and focused on hypervalent iodine mediated C–H amination and copper-catalyzed oxidative cyclopropanation. In 2017, he joined the David Procter group as an EPSRC postdoctoral fellow at the University of Manchester (UK) and was awarded a Marie Curie Fellowship in 2018. In the Procter group, he focused on the development of copper-catalyzed asymmetric multicomponent coupling reactions. In 2020, he moved to Stockholm University (Sweden) to work in the group of Prof. Jan-Erling Bäckvall, currently, he is focusing on iron-catalyzed biomimetic oxidation.

Wei-Jun Kong received his PhD degree from Shanghai Institute of Organic Chemistry, Chinese Academy of Science in 2017 under the supervision of Prof. Jin-Quan Yu. He then joined Prof. Lutz Ackermann's group as a postdoc at Gottingen University. From 2019 to now, he works as a postdoctoral researcher in the group of Prof. Jan-Erling Bäckvall at Stockholm University. His research is mainly focused on transition metal-catalyzed aerobic and electrochemical oxidations in organic synthesis.

Jan-Erling Bäckvall was born in Malung, Sweden, in 1947. He received his PhD from the Royal Institute of Technology, Stockholm, in 1975 with Prof. B. Åkermark. After postdoctoral work (1975-1976) with Prof. K. B. Sharpless, he joined the faculty at the Royal Institute of Technology. He was appointed professor of organic chemistry at Uppsala University in 1986. In 1997, he moved to Stockholm University where he is currently a professor of organic chemistry. He is a member of the Royal Swedish Academy of Sciences, Finnish Academy of Science and Letters, and Academia Europaea. He was Chairman of the Editorial Board of Chemistry—A European Journal 2003–2018 and a Member of the Nobel Committee for Chemistry 2008–2016. His current research interests include transition-metal-catalyzed organic transformations, biomimetic oxidations, and enzyme catalysis.

CHAPTER TWO

Zeolites catalyze selective reactions of large organic molecules

Marta Mon and Antonio Leyva-Pérez*

Instituto de Tecnología Química (UPV–CSIC), Universitat Politècnica de València–Consejo Superior de Investigaciones Científicas Avda. de los Naranjos s/n, Valencia, Spain
*Corresponding author: e-mail address: anleyva@itq.upv.es

Contents

1. Introduction	60
2. Metathesis reaction	62
2.1 Introduction	62
2.2 Metathesis of small alkenes	64
2.3 Metathesis reaction of medium–size molecules	67
3. Hydroaddition reactions to alkenes and alkynes	69
3.1 Introduction	69
3.2 Hydroboration (carbon–boron)	70
3.3 Hydroamination (carbon–nitrogen)	71
3.4 Hydration/hydroalkoxylation (carbon–oxygen)	75
3.5 Hydrophosphination (carbon–phosphorous)	79
3.6 Hydrosilylation (carbon–silicon)	80
3.7 Hydrothiolation (carbon–sulfur)	83
4. Electrocyclization reactions	85
4.1 Introduction	85
4.2 Diels–Alder and Pauson–Khand reactions	85
4.3 Nazarov reaction	86
5. Imination reactions	88
6. Selective alkylation reactions	89
6.1 Alpha–alkylation of ketones	90
6.2 Alkylation of arenes and alcohols	92
7. Conclusions and outlook	96
Acknowledgments	97
References	97
About the authors	102

Advances in Catalysis, Volume 69
ISSN 0360-0564
https://doi.org/10.1016/bs.acat.2021.11.002

Copyright © 2021 Elsevier Inc.
All rights reserved.

59

Abstract

Zeolites are the most used catalysts by weight in chemical industry, but fundamentally for petrochemical processes with small (up to C_6) low functionalized molecules. This chapter deals with recent advances in the use of zeolites as catalysts for selective transformations of relatively large, functionalized molecules, of utility in advanced organic synthesis, fine chemistry and pharma. The variety of zeolites currently available allows to exquisitely control the nature of the active sites, both inside and outside the zeolite channels, to better tune the catalytic activity toward the desired products. It will be shown here how it is not anymore needed that the reagents mandatorily penetrate inside the channels and cavities of the zeolite, and the outer surface and the interparticle space of the zeolites can act as reaction playgrounds. Besides, the zeolite can also act as a macroligand to finely tune the catalytic activity of supported metal ions, either exchanged or impregnated, thus compensating the natural anionic charge of the zeolite. All these features make zeolites suitable catalysts to transform organic molecules of medium and large molecular size into valuable intermediates and products for the organic synthesis community, including industry, with the associated benefits of robustness, recyclability, low toxicity and cheap price of these solid catalysts, already well exploited by the petrochemical manufacturing industry. For comparison purposes, seminal studies with small molecules and other aluminosilicate solid materials will also be commented.

Abbreviations

FAU	faujasite
MFI	pentasil type zeolite
ZSM-5	zeolite socony mobil–5
MOR	mordenite
MTO	methyltrioxorhenium(VII)
USY	ultrastable Y zeolite
DFT	density functional theory
MCM	mobil composition of matter
SBA-15	Santa Barbara amorphous–15
RCM	ring–closing metathesis
NMR	nuclear magnetic resonance

1. Introduction

Organic synthesis is continuously expanding its portfolio of chemical reactions, many of them catalyzed by Brönsted and Lewis acids. Some of these organic reactions show a marked quimio– and regio–selectivity, with a reasonable group tolerance, thus amenable for the construction of large size molecules and late stage modifications. The chemical specificity of these reactions has somehow discouraged researchers to explore acid solids as catalysts, particularly acid micro–structured solids such as zeolites where the diffusion of medium size (>0.7 nm) organic molecules is, in principle, severely restricted. However, advances in the last years have clearly shown that relatively large organic molecules can react with the help of zeolite catalysts, not only because the supercavities of the zeolites can accommodate a variety of organic building blocks, but also because the enhanced diffusion through the pores and the catalytic activity of the external surface and the interparticle mesoporous channels provide alternative molecular mechanisms to carry out the catalytic events.

Zeolites are crystalline aluminosilicate solids, formed by TO_4 tetrahedra ($T = Si$ or Al) connected through the oxygen atoms, giving rise to a network of crossing channels of around 0.5–1 nm porous size and cavities of up to 1.5 nm. When the structure is formed only by SiO_4^{4-}, the solid is neutral, however, the isomorphous substitution of Si^{4+} per Al^{3+} gives rise to an excess of negative charge that is counterbalanced by extra–framework cations maintaining the neutrality of the network. The topological uniqueness of zeolite makes it have a high inner surface, around $500–800$ m^2 g^{-1}, which in combination with the electronically defective but robust aluminosilicate framework, generates a recoverable acid solid with confinement effects. Moreover, zeolites are commercially available in different Si/Al ratios, porous and particle sizes, at affordable prices ($\sim$20 euros/kg), thus allowing their easy implementation in high scale processes. Not in vain, zeolites are the most used catalysts in weight worldwide, due to their recurrent use in huge-volume petrochemistry process.

Fine chemistry is progressively incorporating zeolites in their catalytic toolbox. Seminal studies go back in time to the 90s, where the works of Van Bekkum (1), Davies (2,3), Sheldon (4), Hölderich (5), Jacobs (6,7), Lercher (8) and Corma (9,10), among others, opened the way for the use of zeolites in fine chemical synthesis. These works mainly focused on classical Brönsted–catalyzed organic reactions, such as condensations (11), Friedel–Crafts alkylations (6,7,9), isomerizations (8,10,12) and cyclization reactions (13), and also in oxidation reactions (4,14) by early incorporating metal sites into the zeolitic structure. These studies have arrived to our days with new findings still appearing, since the possibilities of these classical organic reactions are endless. However, a tendency to study more complex reactions can be observed in the open literature, which is the natural way to really incorporate zeolites in general organic chemistry. Nevertheless, the real application of these zeolite–catalyzed processes are mainly applied in the biomass field *(15–17)*, where low regioselective reactions are used, thus it can be said that the research community has not totally embraced yet the goodness of zeolites to selectively catalyze advanced organic reactions.

This chapter deals with the advances reported during the last 10 years in the use of zeolites for five types of acid–catalyzed organic reactions, all of them of readily application in fine chemical and pharma synthesis, where the selectivity and the scope of the reaction are key. These reactions include: (i) metathesis reactions, (ii) hydroaddition reactions to alkenes and alkynes, (iii) electrocyclization reactions, (iv) imination reactions and (v) selective alkylation reactions. These reactions are either carbon–carbon or carbon–heteroatom bond–forming reactions, both types in some cases, thus allow building up a wide array of molecular structures. Thus, the use of recyclable, widely available zeolites for these reactions will provide cheap and environmentally–friendly processes, implementable in–flow, thus fulfilling the sustainable requirements of modern industry and circular economy. When possible, and for the sake of comparison and clarifying, seminal studies with small molecules will also be commented. For the same reasons, other zeolite–type structures such as mesoporous and laminated aluminosilicate materials, will also be incorporated into the discussion.

2. Metathesis reaction
2.1 Introduction

Olefin metathesis consists of the formation of new carbon–carbon bonds by reaction of two olefins that are broken and rearranged. There are specific types of metathesis depending on the substrate involved, as can be seen in Fig. 1A. This reaction, formerly known as olefin disproportionation, was discovered more than 50 years ago and was first applied on an industrial scale at Philips Petroleum Company, where Banks and Bailey found by chance that propylene could be catalytically converted to ethylene and butene using a supported molybdenum or tungsten catalyst (Triolefin Process) *(18)*. In 1971, Hérisson and Chauvin proposed the mechanism using a homogeneous tungsten catalyst, which is widely accepted and involves a [2 + 2] cycloaddition between active transition–metal carbenes and olefins forming metallacyclobutanes as reaction intermediates (Fig. 1B) *(19)*. Since then, the interest in olefin metathesis has been increasing, considering that it is of great importance in the field of organic chemistry and has a crucial impact on the pharmaceutical, petroleum or polymer industries, standing out for the

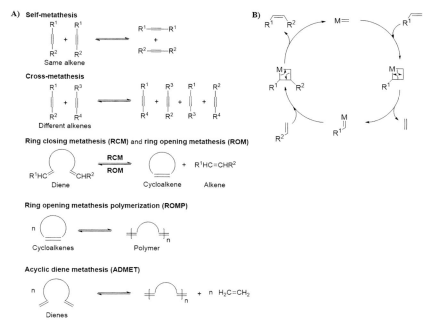

Fig. 1 (A) Schematic representation of metathesis reactions. (B) The general metathesis mechanism proposed by Chauvin.

availability and low price of the starting materials *(20–22)*. In fact, in 2005 the Nobel Prize in chemistry was awarded to Y. Chauvin, R. H. Grubbs, and R. E. Schrock for their contributions made in the study of the reaction mechanism as well as the development of specific and efficient catalysts for this reaction *(23)*.

To carry out this transformation, a great variety of homogeneous catalysts based on transition metal complexes (Ru, Mo, W, Re, Rh, …) have been used, often in combination with other cocatalysts such as Bu$_4$Sn, AlEt$_2$Cl, etc. From an industrial point of view, it is more feasible to use heterogeneous catalysts, so supported metal oxides such as WO$_3$/SiO$_2$, MoO$_3$/Al$_2$O$_3$ or Re$_2$O$_7$/Al$_2$O$_3$ *(24,25)*, have also been used. However, they have shown certain limitations due to the fact that they require high temperatures for their activation and present selectivity problems or intolerance to functionalized olefins. Therefore, there is still work to be done, especially in the search and study of supports that allow the generation of active sites or immobilization of active species for olefin metathesis, avoiding the aforementioned limitations *(26)*.

2.2 Metathesis of small alkenes

At the end of the 80s and beginning of the 90s, it was already known that Re_2O_7/Al_2O_3 was a very active and selective catalyst for olefin metathesis, although the load of Re had to be very high, which is economically and environmentally inconvenient *(27)*. However, the presence of silica in the alumina support increased the activity of the catalyst even with a lower Re loading, due to the increased Brønsted acidity of the support *(28)*. In view of these results, the researchers started using zeolites as supports *(29)*. For instance, Fig. 2 shows that the encapsulation of methyltrioxorhenium (MTO) in zeolite Y with different acidity was carried out, to catalyze the metathesis of 1–hexane *(30)*. In this study it was observed that the metathesis reaction of 1–hexene did not occur with zeolites HY alone, on the contrary, isomerization of the substrate took place. However, doing the reaction in the presence of MTO–zeolites of different acidity, the selectivity toward the metathesis product was higher, although almost no reactivity was observed using a neutral zeolite as support (MTO–NaY). Besides, when MTO or a neutral zeolite (NaY) was used, no reaction occurred. Furthermore, using a HY zeolite of moderate acidity, the yield and selectivity of the metathesis reaction increased, the higher the host load. Therefore, it was demonstrated that the acid sites of the zeolites played a fundamental role on the catalytic activity, since they have an adjustable acidity and allow modifying the load of the encapsulated complex, controlling the yield and selectivity of olefin metathesis.

Given its industrial relevance, one of the olefin metathesis reactions in which more effort has been put to find very efficient catalysts is the

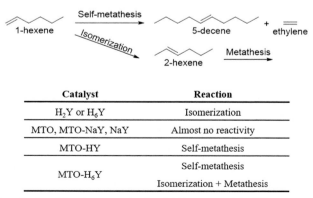

Fig. 2 Results obtained for the metathesis of 1–hexene with different catalysts *(30)*.

cross–metathesis of ethene and 2–butene to obtain propene *(27,31–33)*. This short–chain olefin is suffering a strong global demand, due to the need of the olefin for the production of polymers, such as polypropylene. Conventionally, propene has been obtained as a co–product from steam cracking of naphtha and fluid catalytic cracking. However, other alternatives, such as the reaction discussed here, are being implemented to satisfy their growing demand *(21,34)*. In fact, the production of propene by cross–metathesis of ethene and 2–butene takes place at an industrial level using WO_3/SiO_2 as a catalyst under conditions of high temperature ($>300\ °C$) and pressure ($>20\ bar$). In order to improve these conditions to find an energy–efficient alternative, catalysts based on zeolites are currently being explored. Specifically, a tungsten catalyst (WO_X/USY) has been designed using the cavities of a USY zeolite, in which WO_4 units that are close to unoccupied Brønsted acid sites have been isolated *(32)*. The synergy of these two facts has made this catalyst show very promising results for obtaining propene, since under milder conditions than those used with WO_3/SiO_2 catalyst, a propene production rate approximately 7300 times higher has been obtained, showing good stability. Fig. 3 (top) shows the comparison of the results obtained with both catalysts. The presence of the active WO_4 metathesis sites and unoccupied acid sites, as well as proximity and interaction between them that provides the high activity of the catalyst, have been demonstrated with different experiments and numerous characterization techniques. Recently, another similar catalyst has been designed but, in this case, using rhenium (ReO_x/USY), which has shown an activity and stability even greater than WO_x/USY, that is, 90% propene selectivity at $75\ °C$ and ambient pressure *(33)*. In the ReO_x/USY catalyst, the cooperativity that exists between the isolated active sites of ReO_4 in the cavities of the zeolite and unoccupied H^+ atoms is also observed, facilitating the adsorption of olefins, as well as the cycloaddition $[2+2]$ for the formation of the reaction intermediates. Fig. 3 (bottom) also shows the energy profile calculated by DFT that confirms the importance of the presence of neighboring H^+ atoms. It should be noted that, although it is assumed that the mechanism proposed by Chauvin also occurs when heterogeneous catalysts are used, there are still doubts. Therefore, taking advantage of the advances in characterization techniques, many studies are carried out to understand the nature of the active species in these systems, as well as the relationship of the acidity of the supports, the immobilized metallic precursor, its load, etc., with the activity of the catalyst in metathesis, as in the previous examples.

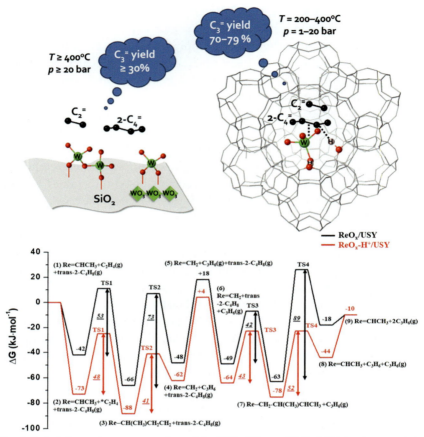

Fig. 3 Top: Schematic representation of the WO_3/SiO_2 and WO_x/USY catalysts and the results obtained for the production of propene (C_3) by cross–metathesis of ethane (C_2) and *trans*–2–butene (2–C_4). Bottom: Energy profiles of the cross–metathesis of ethene and *trans*–2–butene over an active Re=$CHCH_3$ in USY with the absence/presence of a neighboring H^+ atom. *Top: Reprinted with permission from Zhao, P.; Ye, L.; Sun, Z.; Lo, B. T. W.; Woodcock, H.; Huang, C.; Tang, C.; Kirkland, A. I.; Mei, D.; Edman Tsang, S. C. J. Am. Chem. Soc.* **2018**, *140, 6661–6667. Copyright 2018 American Chemical Society. Bottom: Reprinted with permission from Zhao, P.; Ye, L.; Li, G.; Huang, C.; Wu, Si.; Ho, P.-L.; Wang, H.; Yoskamtorn, T.; Sheptyakov, D.; Cibin, G.; Kirkland, A. I.; Tang, C. C.; Zheng, A.; Xue, W.; Mei, D.; Suriye, K.; Edman Tsang, S. C. ACS Catal. **2021**, 11, 3530–3540. Copyright 2021 American Chemical Society.

2.3 Metathesis reaction of medium–size molecules
2.3.1 Alkene–alkene metathesis

Ru complexes, such as Grubbs and Hoveyda–Grubbs alkylidenes, are known to be good catalysts for olefin metathesis. However, until 2015, zeolites had not been used to immobilize Ru metathesis catalyst since it was considered that their pore size was not large enough, unlike mesoporous molecular sieves, such as MCM–41 or SBA–15 *(35)*. However, since there is a wide variety of zeolites, Čejka et al. took advantage of the acidic and structural properties of lamellar zeolites with MWW topology to immobilize Hoveyda–Grubbs type catalysts and use them in olefin metathesis reactions *(36)*. This type of zeolites has a layered structure with two independent pore systems, which are characterized by having active sites with greater accessibility and reactivity *(37,38)*. Specifically, Fig. 4 shows that MCM–22 (a 3D zeolite), MCM–56 (an unilamellar zeolite) and MCM–36 (pillared zeolite) were used to successfully immobilize Ru complexes. Fig. 4 also shows that the ring–closing metathesis (RCM) of (−)–β–citronellene and *N,N*–diallyl–2,2,2–trifluoroacetamide (DAF), the self–metathesis of methyl oleate and the cross–metathesis of methyl oleate with *cis*–3–hexenyl acetate were carried out with these novel catalysts. In this way, they showed that lamellar zeolites were good supports to immobilize Ru complexes and even had higher activity than SBA–15 in the RCM of citronellene and DAF, perhaps due to the better accessibility of the catalytic centers.

2.3.2 Alkene–carbonyl metathesis

The carbonyl–olefin metathesis reaction is a less explored type of olefin metathesis that allows the formation of C—C bonds from an olefin and a carbonyl, rather than two alkenes, as can be seen in Fig. 5. In recent years, advances have been made in this type of reaction, but in all cases using homogeneous catalysts for intramolecular reactions *(39)*. In general, to catalyze this reaction, very acid catalysts are required to increase the reaction rate toward the products *(40)*, therefore there are hardly any example of solid catalysts in the literature *(41)*, since their high acidity can trigger other types of more favorable reactions in an acid medium *(42)*.

Recently, it has been studied the use of the affordable and widely available aluminosilicate montmorillonite K10 to catalyze intermolecular metathesis reactions between aromatic ketones and aldehydes **1a** with vinyl esters **2a**, as shown in Fig. 6 *(43)*. It was demonstrated that the strategy of using vinyl ethers instead of regular alkenes favors the action of the catalyst,

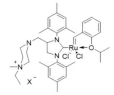

Fig. 4 (A) Artistic representation of some lamellar zeolites. (B) Ru complexes immobilized on zeolitic supports. (C) Olefin metathesis reactions carried out using lamellar zeolites to immobilize Hoveyda–Grubbs type catalysts.

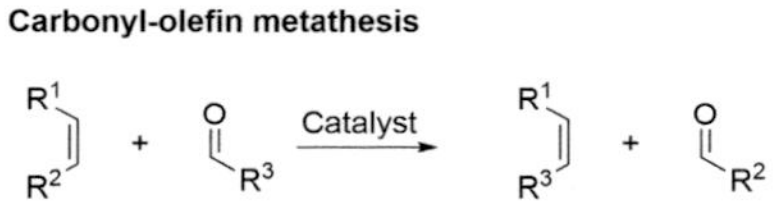

Fig. 5 Schematic representation of carbonyl–olefin metathesis.

Zeolites catalyze selective reactions of large organic molecules 69

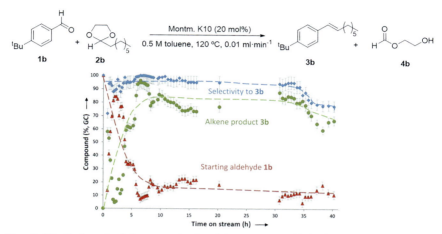

Fig. 6 (A) Carbonyl–olefin metathesis of ketones or aldehydes with vinyl ethers. (B) Kinetic profile for metathesis reaction between aryl aldehyde **1b** and the in situ formed vinyl ether from **2b** catalyzed by montmorillonite K10, in a fixed–bed tubular reactor (43).

by shifting the equilibrium toward the products, which allows obtaining stereoselectively *trans* alkenes **3a** with high yields, on a multi–gram scale and in flow for reagents **1b** and **2b**. These results represent an advance for the implementation of carbonyl–olefin metathesis reactions using heterogeneous catalysts, also in flow.

3. Hydroaddition reactions to alkenes and alkynes

3.1 Introduction

Hydroaddition reactions consist of the direct addition of X–H groups (where X=heteroatom) to unsaturated bonds of alkenes or alkynes. Depending on the nucleophilic group (X–H) which is added to the double or triple bond, the reactions has a specific name such as: hydroboration (B—H), hydroamination (N—H), hydration (H$_2$O), hydroalkoxylation (O—H), hydrophosphination (P—H), hydrosilylation (Si—H) or hydrothiolation (S—H) (44–46). In the case of alkenes, the final product is an alkyl functionalized molecule, while in the case of alkynes, the final product is a vinyl functionalized compound, and in both cases the atom economy is 100%. The regioselectivity can be either Markovnikov, i.e., the addition occurs on the more substituted carbon atom, or anti–Markovnikov, i.e., the addition occurs in the less substituted carbon atom, for example in the terminal carbon atom of terminal alkenes and alkynes. The added functional group (nucleophile) often marks the preponderance of one or another regioselectivity, although the catalyst can also play a role. Another factor to consider for not only the regioselectivity but also for the overall reactivity is the molecularity of the reaction, i.e., intra– or intermolecular. When both reacting functionalities are in the same molecule, at a distance to generate five or six membered cycles, the intramolecular reaction is of course much more favored and the regioselectivity outcome is often driven by the stability of the final product. Besides, the obtained hydroaddition product is sometimes unstable and, in order to avoid decomposition, it is in–situ further functionalized to a more stable product, a typical example being the oxidation of alkyl boronates to the corresponding alcohols. All these features make the hydroaddition of different functional groups to carbon–carbon unsaturated bonds an extraordinary tool to obtain new bonds of utility in organic synthesis.

Zeolites have been employed to catalyze these transformations. In some cases, a supported metal is required to catalyze the reaction, because the zeolite itself cannot mediate in the transformation. In other cases, simple zeolites catalyze the hydroaddition reaction efficiently, and they can be recycled. The fact that alkenes and alkynes are rigid planar molecules can help in the diffusion into the channels, however, if the substituents around the unsaturated double bond are big, this can result in the contrary, since the rigid bulky molecule will not diffuse well through the pores. In this case, the outer zeolite surface may play a catalytic role. It will be found below that some catalytic solids perform efficiently different hydroaddition reactions, since the activation of the alkene or alkyne is common to these processes.

3.2 Hydroboration (carbon–boron)

The hydroboration of alkenes and alkynes, is not necessarily catalyzed, and it can occur spontaneously for certain substrates (47). However, for the most of alkenes and alkynes studied and particularly for certain boranes, a catalyst is highly recommended in order to circumvent excessively long reaction times or partial selectivity. A typical metal catalyst for this reaction is copper, but many other transition metals have been reported to catalyze the reaction (48).

Since Brönsted acidity or basicity does not catalyze the reaction, zeolites have been employed as supports for organometallic complexes that catalyze the reaction. One early example was the use of chiral Rh(I) complexes, with N,N'–, and N,P–ligands, supported on silica and modified USY–zeolite. The technique to immobilize the complex is very common in silicas but no so common on zeolites, and consists in the previous functionalization of the solid surface with silane reagents to give a very reactive functional group (i.e., isocyanate) ready to react with a pending group of the functionalized complex (i.e., alcohol), which must not participate during the catalytic process. In this way, the reaction of styrene with catechol borane in THF, in the presence of 1 mol% of the supported rhodium catalyst, gave 1–phenylethanol as the single product after oxidation of the intermediate alkyl borane. Unfortunately, no enantioselectivity was found (49). However, a later work by other authors (50) showed that other optically active rhodium(I) complexes supported on montmorillonite K–10 and sodium bentonite, through an impregnation instead that a covalent immobilization procedure, gave the enantioselective hydroboration of vinylarenes, with good recyclability. In all the cases reported, the product followed the anti–Markovnikov regioselectivity, typical for hydroboration reactions since the boron atom does not act as a classical nucleophile and it is more stable when engaging the less substituted carbon atom. Fig. 7 shows the structure of the supported complexes and their different interaction with the two solid silicates, where the cationic complexes substitutes its anion $X = BF_4^-, PF_6^-, \ldots$ by the anionic charge of the solid framework. In this way, the supported Rh(I) complex does not leach from the support under reaction conditions and exhibit a comparable activity and selectivity to the free catalyst, as also shown in Fig. 7.

3.3 Hydroamination (carbon–nitrogen)

The hydroamination of alkenes is catalyzed by Brönsted acids, in contrast with the hydroboration reaction. Thus, zeolites are in principle good candidates to catalyze the reaction. Indeed, early studies by the group of Lercher at the beginning of this century found that the simple acid zeolite H–Beta catalyzed efficently the hydroamination of cyclohexadiene with aromatic amines (51). Curiously, this report was preceeded by the corresponding intra– and intermolecular hydroamination of alkynes, where

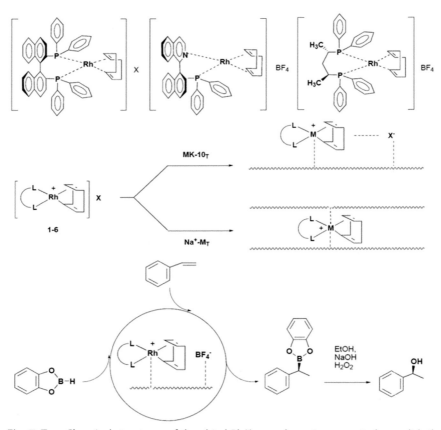

Fig. 7 Top: Chemical structures of the chiral Rh(I) complexes impregnated on solid silicates. Middle: The proposed complex–surface interaction. Bottom: hydroboration of styrene catalyzed by the solid catalysts. *Adapted from Segarra, A. M.; Guerrero, R.; Claver, C.; Fernandez, E. Chem. Eur. J.* **2003**, *9 (1), 191–200s.*

the same zeolite exchanged with Zn^{2+} was used as a catalyst for the reaction *(52–54)*. Other zeolite–type materials exchanged with metals, i.e., Cu^{2+}, were also employed for the same transformation *(55)*. In all these cases, the transformation follows the Markovnikov rule, since the amine acts as a classical nucleophile.

The scope of alkenes and amines was rapidly expanded to others, such as acrylates, aryl amines and imidazoles *(56–58)*, with structure–activity realtionships studies of the zeolite and the reagents *(59)*. It was also found in these studies that Beta and mordenite zeolites, exchanged or not with scandium triflate, were able to catalyze the hydroamination of styrenes as a function of the strength of the acid site and its accessibility, while the selectivity appeared to be controlled by the Lewis/Brönsted type of acidity. Indeed, Lewis acidity directed to the formation of the Markovnikov adduct while Bronsted acidity directed to the anti–Markovnikov product. Notice that the addition to acrylates, although it is formally a hydroamination of the alkene, it is better known as a Michael addition reaction, which is easier to proceed. Indeed, simple Na—Y zeolite is able to catalyze the hydroamination of α,β–unsaturated compounds with aromatic amines (60).

As it occurs for the hydroboration reaction, Rh–supported zeolites have also shown catalytic activity during the hydroamination reaction and, in particular, Rh exchanged titanosilicates are efficient, stable and reusable catalysts for the hydroaminomethylation reaction of 1–hexene and pyrrolidine (61). However, later studies showed that, if conveniently prepared, bare titanosilicates and zirconosilicate in nanocrystalline form catalyzed the amines to methyl acrylate, inclusing structure to catalytic activity relationship studies (62). ZSM–5 zeolites have also shown catalytic activity for the hydroamination reaction (63). In this work, the catalytic activity of the ZSM–5 materials was compared with that of the conventional ZSM–5 and amorphous mesoporous aluminosilicate Al–MCM–41.

Thus, the following challenge was to catalyze the reaction in pure Lewis zeolite catalysts, without any Brönsted acidity (64). This was achieved by the group of Corma employing a bifunctional Au—Sn organic–inorganic catalyst, where gold(III) complexes heterogenized on the surface of a Sn–containing MCM–41 were used as efficient and recyclable catalysts for the hydroamination reaction (65). More recently, the judiously employment of combined acid/redox sites in zeolites allowed to get the hydroamination reaction (66). Later, Pt^{2+} exchanged zeolites were prepared to catalyze the hydroamination reaction (67). In this work, it was found that the typical Pt and Pd salts employed as catalysts for this transformation (and related ones) in homogeneous phase are indeed insoluble and only a small fraction of dettached metal ions are active for the hydroamination reaction, as shown in Fig. 8. Thus, it was envisioned that a good dispersion of metal cations along the zeolite surface will render a very efficient catalyst for the hydroamination reaction.

The solid Pt(II)–zeolite catalyst was prepared by exchanging soluble $[Pt(NH_3)_4]^{2+}$ with the interchangable Na^+ cations of zeolite NaY. Pt(II) was easily introduced in NaY zeolite in this way, to obtain a $[Pt(NH_3)_4]^{2+}$ –exchanged zeolite. This solid was not active for the reaction, since the NH_3 ligands inhibited the coordination of the reagents, i.e., alkyne and

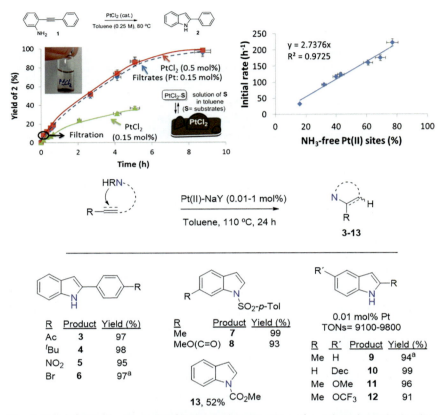

Fig. 8 Top left: Yield – time plot for the hydroamination of o–(phenylacetylen)aniline **1** to give indole **2** catalyzed by PtCl$_2$ under the indicated reaction conditions. Each point is an average of three different measurements, and bars represent an estimated error of 5%. The top left inset shows a photograph of the PtCl$_2$ remaining, and the bottom right inset illustrates the equilibrium–controlled dissolution of PtCl$_2$ with the substrates. Top right: Initial conversion rates for the hydroamination of **1**, at 110 °C, catalyzed by different Pt(II) – Y zeolites (1 mol%) where the number of NH$_3$–free Pt(II) sites has been varied. Error bars account for 5% uncertainty. Bottom: Hydroadditions to alkynes catalyzed by Pt(II) – NaY, to give products **3–13**. Isolated yields, 1 mol% of Pt otherwise indicated. TON: turnover number. [a] Pt(II) – NaY reused 3 times. Adapted from Rubio-Marqués, P.; Rivero-Crespo, M. A.; Leyva-Pérez, A.; Corma, A. *J. Am. Chem. Soc.* 2015, 137, 11832–118379s.

amine, to the Pt(II) active site. However, a simple calcination of the catalyst at different temperatures allowed to release the volatile NH$_3$ ligands, the anionic framework now acting as a macroligand for Pt(II). In this way, different degrees of deamination of the Pt(II) site could be obtanied, and Fig. 8

also shows that the higher the deamination, the better the catalytic activity. This in in line with the observed detachment of Pt(II) cations from $PtCl_2$ by the very same alkyne and amine reagents, needed to form the catalyically active species in solution. With the new solid catalysts in hand, a variety of aminoalkynes could be intramolecularly hydroaminated in high yields and selectivity, to give products **3–13**, reaching turnover numbers near 10,000, as it shown in Fig. 8.

Remarkably, it has been recently reported that the reactivity of the aryl amine can be shifted form the amine to the aryl group, thus giving hydroarylation instead than hydroamination products *(68)*. USY zeolite was used as a catalyst for this unusual transformation, where different alkenes and aromatic amines with various functional groups were smoothly converted into the corresponding products with high regioselectivity, after a Lewis acid–promoted Hofmann–Martius rearrangement of the hydro-amination toward the hydroarylation products. Authors propose that the weak acid sites of the zeolite play a key role in forming hydroarylation products, and the solid catalyst could be reused at least 10 times without deactivation.

3.4 Hydration/hydroalkoxylation (carbon–oxygen)

The hydration and hydroalkoxylation reactions consist in the addition of water or alcohols, respectively, to unsaturated carbon–carbon double bonds. Here, a huge difference in reactivity can be found between alkenes and alkynes. In contrast to the hydroboration reaction, where alkenes are easier to react (even uncatalyzed) than alkynes, the hydration and the hydro-alkoxylation of alkynes is much easier than alkenes, since the corresponding products ketone and enol ether, respectively, are much more stable. In some cases, these reactions are related and the hydroalkoxylation often precedes the hydration reaction, since the enol intermediate can easily be hydrated to the corresponding ketone (the hydration product) *(69,70)*. However, both products ketone (hydration) and enol ether (hydroalkoxylation) have been independently obtained with the appropriate zeolite catalyst, as it will be seen below. As it occurs for the hydroamination reaction, the oxygen nucleophiles here act classically to give nearly exclusively the Markovnikov product. A very common variant of the hydration reaction is the Meyer–Schuster rearrangement of propargylic alcohols to give α,β–unsaturated aldehydes and ketones, which will be also treated in this section.

3.4.1 Hydration reaction

The hydration of alkynes catalyzed by zeolites was early reported in 1989 *(71)*. In this seminal work, different Y faujasite zeolites, including HY, CeY, HgY and CdY, catalyzed the hydration of alkynes to ketones at 200 °C. This transformation was then expanded to α–alkynols, to give α,β–unsaturated ketones over zeolite catalysts, following the Meyer–Schuster rearrangement *(72)*. Later, the proton H–Beta zeolite was used as an active catalyst, under solvent–free conditions, for the hydration of alkynes *(73,74)*. The solid catalyst was recyclable and showed the typical Meyer–Schuster rearrangement in propargylic aryl alcohols to give α,β–unsaturated aldehydes in excellent yields.

Fig. 9 shows that a comprehensive study with 24 types of heterogeneous and homogeneous catalysts revealed that the best catalyst for the hydration reaction of diphenylacetylene **14** to ketone **15** is a H–Beta zeolite with a

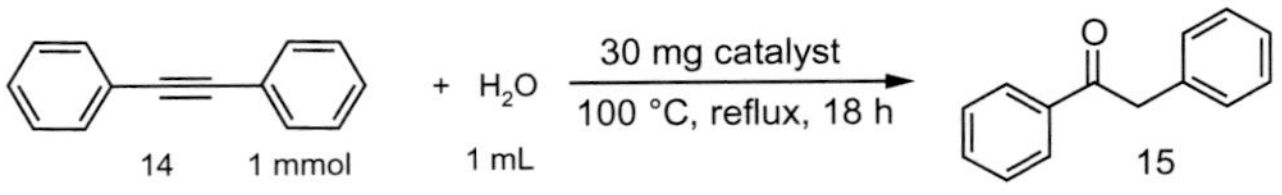

Entry	Catalyst	15 yield (%)[a]
1	none	0
2	HZSM5-11	2
3	HZSM5-75	11
4	HZSM5-150	15
5	Hβ-12.5	44
6	Hβ-20	51
7	Hβ-75	98
8	Hβ-255	11
9	HY-50	15
10	HMOR-45	78
11	SiO_2-Al_2O_3	3
12	SiO_2	16
13	ZrO_2	2
14	TiO_2	6
15	SnO_2	4
16	CeO_2	1
17	Nb_2O_5	5
18	Niobic acid	43
19	$Cs_{25}H_{0.5}PW_{12}O_{40}$	13
20	Mont. K10	55
21	Nafion-SiO_2	76
22	Amberlyst-15	65
23	H_2SO_4	43
24	PTSA	47
25	$Sc(OTf)_3$	36

[a]GC yield.

Fig. 9 Catalyst screening for the hydration reaction of diphenylacetylene **14** to ketone **15**. *Adapted from Sultana Poly, S.; Hakim Siddiki, S. M. A.; Touchy, A. S.; Yasumura, S.; Toyao, T.; Maeno, Z.; Shimizu, K.-i. J. Catal. 2018, 368, 145–154.*

relatively high Si/Al ratio (Si/Al = 75) *(75)*. The solid catalyst also showed wide substrate scope, good reusability, and applicability in Gram–scale synthesis.

Later works have shown that the hydration of alkynes can be accomplished with other more advanced, but still simple, zeolite materials. For instance, zeolite Y nanoparticle assemblies showed high activity in the direct hydration of terminal alkynes, being reusable for at least six times *(76)*. The strong acid assembly with a micro–*meso*–macroporous structure showed higher catalytic activity than acidic mesoporous zeolite ZSM–5 and Beta catalysts. These findings were assigned to the fact that the micro–meso–macroporous structure benefits mass transfer and access of the reagents to the strongly acidic sites.

3.4.2 Hydroalkoxylation reaction

The hydroalkoxylation of alkenes has curiously been reported in both the intermolecular and the intramolecular fashion, in similar times. The intramolecular hydroalkoxylation of several alkenes was experimental and computationally studied with different zeolites, including MFI, FAU and BEA types *(77)*. The best catalytic performance was found for zeolites with an optimum concentration of Bronsted acid sites (ca. 0.2 mmol g^{-1}) and a minimum concentration of Lewis acid sites. The intramolecular hydroalkoxylation of alkynes have been studied with the Pt(II)–zeolite catalysts mentioned above for the hydroamination reaction *(67)*.

The intermolecular reaction of 1–hexene with 1–propanol and 1–butanol, which are obtained from biomass sources, has been accomplished with zeolite beta as a catalyst in a liquid phase continuous–flow process, over a fixed bed of the solid catalyst *(78)*. The zeolite was stable for at least 14 h on stream, with over 90% selectivity for 2–propoxyhexane. The intermolecular hydroalkoxylation of α,β–unsaturated esters with different aliphatic and aromatic alcohols has also been investigated with pollucite nanoparticles, using both thermal and microwave–assisted methods *(79)*.

All the above works based the catalytic activity on the control of the zeolite active acid sites. However, Fig. 10 shows that carbocations **17** and **18** are plausible intermediate during the hydration/hydroalkoxylation reactions, which is indeed the basis for the Meyer–Schuster rearrangements in propargyl alcohols such as **16** *(80)*. Considering this, it was envisioned that the stabilization of the intermediate carbocations by the zeolite framework, rather than the activation of the substrate, could lead to a better catalytic

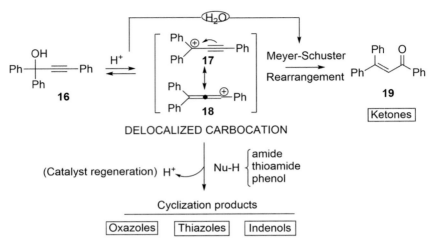

Fig. 10 Formation of a delocalized carbocations **17** and **18** from propargyl alcohol **16** with a proton, and catalytic addition of amides, thioamides, and phenols to give oxazoles, thiazoles, and indenols, respectively. In the absence of any other nucleophile water often re–enters to give the Meyer–Schuster rearrangement to ketone **19**. Adapted from Cabrero-Antonino, J. R.; Leyva-Pérez, A.; Corma, A. Angew. Chem. Int. Ed. 2015, 54, 5658–5661.

activity of the zeolite toward hydroalkoxylation products, thus avoiding the formation of the Meyer–Schuster product **19**.

Indeed, Fig. 11 shows that the Meyer–Schuster rearrangement of propargyl alcohols **16** catalyzed by different soluble sulfonic acids and zeolite H–USY gives a straight line for the soluble acids, when plotted the activation energy of the reaction vs the acidity strength, in either pKa or Ho values. In striking contrast, the activation energy of the Meyer–Schuster rearrangement dramatically decreases for the zeolite, showcasing the beneficial effect of stabilizing the carbocation by the anionic framework of the zeolite.

Fig. 12 shows that zeolite H–USY was able to catalyze the intramolecular cyclization of medium–size propargyl amides, generated in–situ from the corresponding propargyl alcohols and amides, to give the corresponding oxazole heterocycles **20–25** in good yields and without needing a particularly high acidity.

Notice that the size of these molecules, including reagents and products, is not amenable to diffuse through the zeolite channels, thus the most of the catalytic activity here can be ascribed to the outer surface of the zeolite, including the interparticle space. Indeed, a nanocrystalline H–Beta zeolite gave a better catalytic performance compared with the regular H–Beta

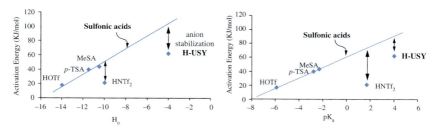

Fig. 11 Activation energy–acidity values (H_0, left; pK_a, right) plot for different acids and H–USY zeolite. The activation energies of the reaction are calculated from the initial rate of the Meyer–Schuster rearrangement of **16**, by in–situ NMR measurements. High catalytic loadings of H–USY (120 wt%) and $HNTf_2$ assures rapid formation of the ketone. *Adapted from Cabrero-Antonino, J. R.; Leyva-Pérez, A.; Corma, A. Angew. Chem. Int. Ed.* **2015**, *54, 5658–5661*.

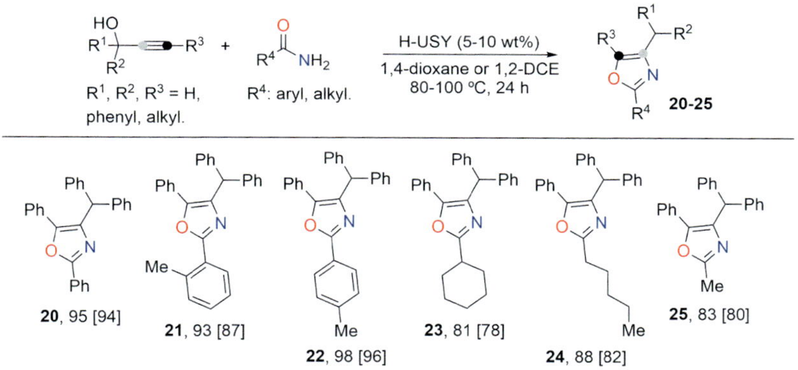

Fig. 12 Scope of the cyclization reaction between substituted propargyl alcohols and amides, catalyzed by H–USY zeolite (Si/Al = 15). GC and isolated yields [%]. *Adapted from Cabrero-Antonino, J. R.; Leyva-Pérez, A.; Corma, A. Angew. Chem. Int. Ed.* **2015**, *54, 5658–5661*.

zeolite, due to the higher outer surface and closer interparticle space of the former. Biological studies show that some of the products obtained here present significant inhibition activity against colon cancer cells, illustrating the new possibilities of zeolites to prepare complex organic molecules with application in pharma *(80)*.

3.5 Hydrophosphination (carbon–phosphorous)

The addition of phosphines across carbon–carbon double and triple bonds is not generally catalyzed by protons but requires a metal catalyst, as occurs with boron nucleophiles *(81,82)*. Thus, it is not surprising that the few examples of zeolite–catalyzed hydrophosphination reactions involve Rh–supported zeolites. For instance, a diphosphino–functionalized MCM–41 was used to immobilize a Rh complex, and this recyclable solid catalyst showed a good catalytic activity for the hydrophosphination of terminal alkynes with triphenylphosphine oxide, being the solid catalyst recyclable 10 times *(83,84)*.

3.6 Hydrosilylation (carbon–silicon)

Hydrosilation is a very important reaction from an industrial point of view to obtain organo–silicon compounds, in which homogeneous Pt catalysts in part–per–million amounts are usually used *(85)*. The fact that the soluble Pt catalyst is not removed after the reaction and that the produced orga-nosilanes are present in industrial silicones in multi–tonnes per year, has led to a worldwide spread of this inherently toxic metal. Thus, the search for more benign hydrosilylation catalysts, particualry recyclable solids, with similar catalytic activity to soluble Pt, is a topic of high interest *(86)*. Indeed, there are not many examples of truly recyclable heterogeneous catalysts with good results to carry out this reaction *(87)*. Consequently, the search for sus-tainable and recoverable solid metal catalysts still continues, and despite research on heterogeneous catalysts has increased in recent years, the resulting solid catalysts are not stable and therefore recyclable. Here, zeolites may be a solution when used as a support to stabilize the active species in their channels *(88)*.

Most of the solid catalysts employed for the hydrosylilation of alkenes in the recent literature are conformed by precious metal species on alumino-silicate type materials, but not zeolites. For instance, atomically dispersed single Pt atoms on alumina nanorods (Pt/NR–Al_2O_3), formed by impreg-nation of tiny amounts of soluble Pt species on the high surface alumina material, were efficiently employed as a catalyst for the hydrosilylation of a plethora of alkenes, to give the corresponding alkyl silanes, including some with commercial relevance *(89)*. A similar approach, but using titania as a support, has also been reported *(90)*, and other supperts such as ceria were used to immobilize catalytically active gold species *(91)*.

Particularly common has been the use of MCM–41 as a support for cat-alytic active species of the hyddrosilylation reactions. For instance, a Ru complex ([CpRuBz]PF_6 (Bz $= C_6H_6$)) has been immobilized onto amino–functionalized MCM–41 (MCM–41–NH_2) through amide bond formation *(92)*. Another catalytically active Cu(I)–NHC species has also been

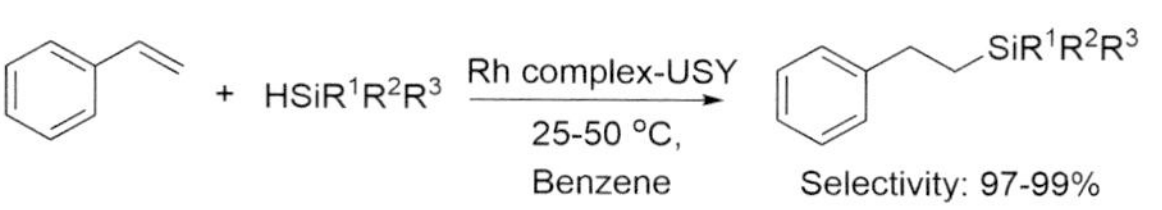

Fig. 13 Hydrosilylation reaction of styrene with different silanes catalyzed by Rh complexes anchored to USY–zeolites.

incorporated in MCM–41 *(93)*. The same material above employed for the hydrophosphination reaction, i.e., a Rh complex stabilized on phosphino fucntionalized MCM–41 (MCM–41–2P–RhCl$_3$) was also employed as a catalyst for this reaction *(94)*, however, despite operating under mild conditions, Rh complexes often lead to other secondary reactions such as the hydrogenation and/or dehydrogenating silylation of olefins.

The use of zeolites to incorporate Rh complexes has also been reported *(95)* and Fig. 13 shows that, in contrast to the mesoporous supports, these Rh supporting zeolites avoid significantly the undesired by–reactions, thus the hydrosilylation of styrene with different silanes proceeded in the presence of rhodium complexes anchored on a modified USY–zeolites with unusual regioselectivity to give 2–phenylethylsilanes in excellent yields. This catalyst showed greater stability and selectivity with respect to its homogeneous counterparts. In addition, no metal leaching was observed and the solid catalyst could be perfomed during five cycles without losing activity.

In view of this, zeolites with the Pt single atoms directly incorporated into the framework, as shown above for the hydroamination reaction *(67)*, were used *(96)*. Fig. 14 shows that the typical NaY zeolite (Pt$_1$/NaY) was extremely efficient for the hydrosilylation of styrene **26** with silane **27**, to give the corresponding alkyl silane **28**, with the typical anti–Markovnikov selectivity (the Si atom behaves here as the B atom, as a non–classical nucleophile). The amine ligands of the Pt$_1$/NaY catalysts must be conveniently removed by calcination at 200 °C (see above), in order to achieve an optimum catalytic activity (expressed as turnover frequency), even better than the industrial soluble Pt catalysts. A related microstructured metal organic framework (MOF), also containing the Pt$_1$ species, was similarly effective for the reaction (entry 8), which confirms the role of Pt$_1$ as the single active species for the hydrosilylation reaction. Indeed, a variety of alkenes and silanes could be coupled in the presence of this solid catalyst, as also shown in Fig. 14, to give products **28–36**.

The Pt$_1$/NaY solid catalyst was not only effective for alkenes but also for alkynes, with the remarkable ability of swtiching the typical regioselectivity

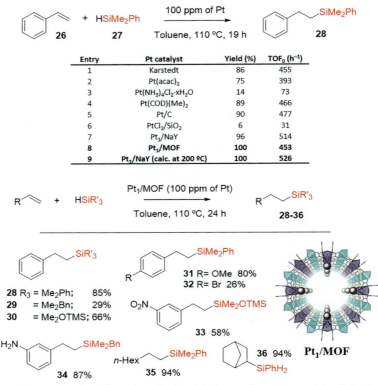

Fig. 14 Top: Results for the hydrosilylation of styrene **26** (0.5 M) with dimethylphenylsilane **27** (1.2 equivalents) to give (2–phenylethyl)dimethylphenylsilane **28** with different soluble and solid Pt catalysts (100 ppm); acac = acetylacetonate, COD = cyclooctadiene. GC yields using *n*–dodecane as an external standard. Bottom: scope for the hydrosilylation of terminal alkenes catalyzed by Pt$_1$–MOF (NMR yields using dibromomethane as an internal standard), and the single crystal X–ray diffractogram of the catalyst (Pt atoms, gray balls). Adapted from Rivero-Crespo, M.; Oliver-Meseguer, J.; Kapłońska, K.; Kuśtrowski, P.; Pardo, E.; Cerón-Carrasco, J. P.; Leyva-Pérez, A. Chem. Sci. **2020**, 11 (31), 8113–8124.

for hydrosylilation reactions from the anti– to the Markovnikov, adducts, as shown in Fig. 15 (96). Here, different alkynes and silanes enage to give the corresponding α–vinylsilanes **37–48** in the presence of the zeolite catalyst, with complete conversions and good selectivities toward the Markovnikov product. Pt$_1$/NaY did not show significant leaching and could be reused without loss of activity.

Experimental and computational studies together with an *ad–hoc* graphical method showed that the hydroaddition of silanes to alkenes and alkynes

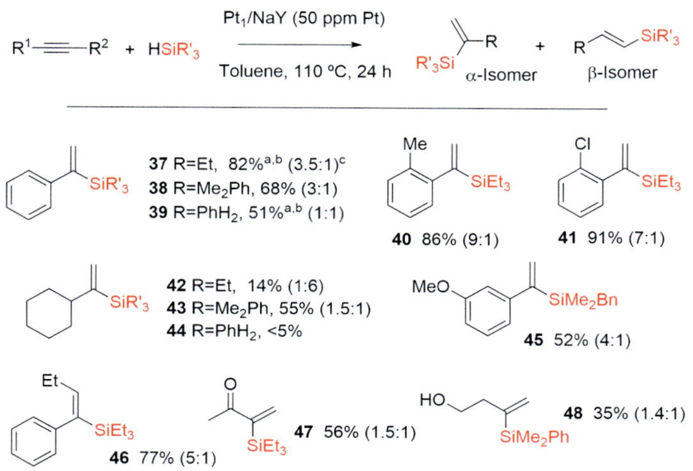

Fig. 15 Hydrosilylation reaction of different alkynes and silanes catalyzed by platinum single atoms confined in NaY zeolite, to give products **37–48** with Markovnikov selectivity. *Adapted from Rivero-Crespo, M.; Oliver-Meseguer, J.; Kapłońska, K.; Kuśtrowski, P.; Pardo, E.; Cerón-Carrasco, J. P.; Leyva-Pérez, A.* Chem. Sci. **2020**, 11 *(31), 8113–8124.*

proceeds either through Pt single atoms or Pt–Si–H clusters of 3–5 atoms (what was called metal(oid) association), which decreased the energy of the transition state and direct the regioselectivity of the reaction *(96)*. Therefore, the use of Pt supported zeolites is one of the few efficient processes avialble where a recyclable and tunable solid catalyst seek to be used for the hydrosilylation of alkenes and alkynes.

3.7 Hydrothiolation (carbon–sulfur)

The addition of thiols to double bonds gives rise to two different products (Markovnikov and anti–Markovnikov) since the reaction can take place in two different ways, that is, by free radicals obtaining the anti–Markovnikov regioisomer, or by an electrophilic mechanism, where the nucleophile is added to the carbon which better stabilizes the positive charge of the intermediate (more substituted carbon atom), obtaining the Markovnikov regioisomer. This behavior differs from oxygen nucleophiles, where the Markovnikov product is always predominant, due to the soft nature of the S atom which allows the stabilization of radicals. The free radical reaction is known to proceed easily under mild conditions in the presence of a radical initiator, even light, but it is anyway interesting to develop heterogeneous catalysts for this reaction. However, electrophilic addition only takes place

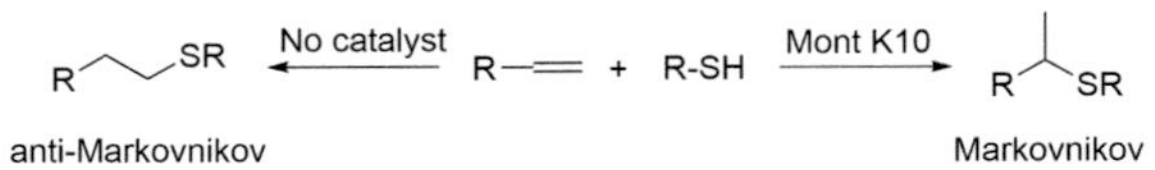

Fig. 16 Hydrothiolation of alkenes in the presence or not of montmorillonite–K10.

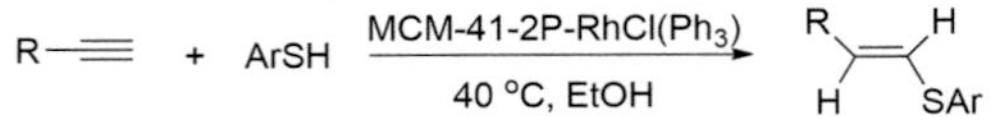

Fig. 17 Hydrothiolation of alkenes with the MCM–41–2P–RhCl(PPh₃) catalyst.

when stoichiometric amounts of Brønsted acids (H_2SO_4, p–toluensulfonic acid or $HClO_4$) or Lewis acids ($AlCl_3$, $TiCl_4$ or BF_3) are used. There are few examples of catalytic systems for the Markovnikov addition of thiols to alkenes, among which are: $In(OTf)_3$, complexes of Fe^{III}, compounds of Au(I) and Au(III) or Montmorillonite–K10 *(97)*. Fig. 16 shows that the hydrothiolation of alkenes with thiols or thioacids can be carried out in the presence and absence of the solid catalyst montmorillonite K10 (Mont K10), observing that, in its presence, the Markovnikov product was obtained and, in its absence, the anti–Markovnikov product was obtained *(98)*.

Not many transition metals have been used to catalyze reactions of organic sulfur compounds as these are often considered incompatible with metals, as they can bind strongly to the transition metal and poison the catalyst. The Ogawa and Love groups *(99, 100)* reported that $RhCl(PPh_3)_3$ is an excellent catalysts for the anti–Markovnikov addition of thiols to terminal alkynes, obtaining (*E*)–vinyl sulfides. However, the use of heterogeneous catalysts for hydrothiolation of alkenes or alkynes is not well explored. Motivated by the results of Ogawa and Love, in 2012 the first solid catalyst was reported to carry out the hydrothiolation of terminal alkynes with thiols *(101)*. Fig. 17 shows that the catalyst (MCM–41–2P–RhCl(PPh₃)), already used above for the hydrophosphination reaction and prepared by anchoring the $RhCl(PPh_3)_3$ complex to diphosphino–functionalized mesoporous material MCM–41, operates under mild conditions and showed similar catalytic activity to the $RhCl(PPh_3)_3$ complex for regio– and stereoselective synthesis of anti–Markovnikov isomers, i.e., (E)–vinyl sulfides. Furthermore, it was shown that the catalyst could be reused. The synthesis of vinyl and alkyl sulfides via hydrothiolation of alkynes and electron–deficient olefins was also accomplished with and heterogenized gold complexes catalysts, already commented above for the hydrosilylation reactions *(102)*.

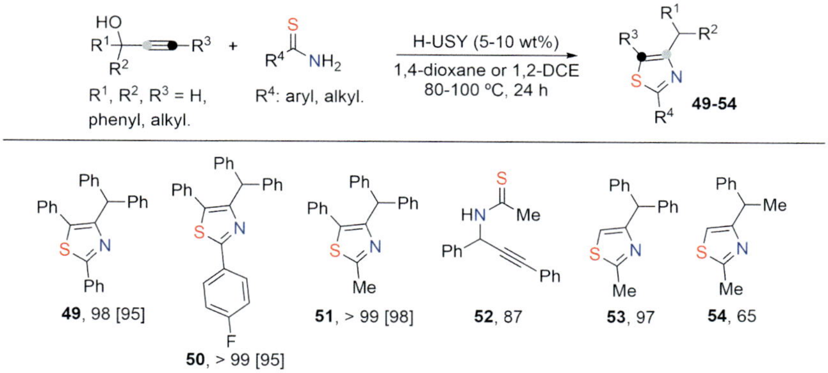

Fig. 18 Intramolecular hydrothiolation of in–situ formed propargyl thioamides, to give thioimidazoles **49–54**, catalyzed by H–USY zeolite (Si/Al = 15). GC and isolated yields [%]. Adapted from Cabrero-Antonino, J. R.; Leyva-Pérez, A.; Corma, A. Angew. Chem. Int. Ed. **2015**, 54, 5658–5661.

Other example of zeolite catalysts for the intramolecular hydrothiolation reaction of alkynes, for the synthesis of pharma useful compounds, is also the above commneted catalytic system where carbocations of propargyl alcohols are stabilized (80). Fig. 18 shows that different thioimidazoles **49–54** were obtanied in high yields with H–USY as a catalyst, from the in–situ reaction of the corresponding amides and propargyl alcohols.

4. Electrocyclization reactions
4.1 Introduction

Electrocyclization reactions involve a series of organic reactions where the electrons move either intra– or intermolecularly to break and form the same number (and nature) of bonds in the reagents and products. Three representative examples of this kind, which cover most of the rest of the group, are the Diels–Alder, Pauson–Khand and Nazarov reactions. We will treat only these reactions in this section regarding the use of zeolites as catalysts, since these reactions are perhaps the most representative reactions in this group.

4.2 Diels–Alder and Pauson–Khand reactions

The Diels Alder reaction is described as a (3) reaction whereas the breaking of three double carbon–carbon bonds to form a new double bond and two new single carbon–carbon bonds, occurs. This reaction is well–known to occur with acid solids and in confined spaces, thus zeolites were early candidates to catalyze this reaction, and it is common to find this reaction in reviews and books on zeolite–catalyzed organic reactions *(103)*. However, recent examples deserve to be commented here.

Roman–Leshkov et al. recently reviewed the use of zeolites as Lewis acid catalysts for carbon–carbon bond formation *(104)*, and they found that the use of zeolites, with their hydrophobic cavities and isolated framework metal sites, can be considered as enzyme mimics, and they have been key to discover new Diels–Alder cycloaddition or dehydration routes, both experimental and computationally. In other recent review, an overview of the heterogeneous Diels–Alder reaction for the manufacturing of lignocellulosic biomass–derived aromatic compounds, employing zeolite catalysts, has been carried out *(105)*.

The Pauson–Khand reaction is a powerful method for the construction of cyclopentenones, however, it requires of metal–carbon monoxide reagents (often Co_2CO_8) or a carbon monoxide atmosphere to be completed. Indeed, the first heterogeneous Pauson–Khand catalyst claimed was composed of Co supported on mesoporous silica, active for the intramolcular Pauson–Khand cyclization of enynes *(106)*. Intermolcelular reactions also worked but in lower yields. Silica SBA–15 was used as support under 20 atm of CO in THF, and the supported cobalt catalyst was stable to air and recyclable. Besides, amorphous silica–supported cobalt catalysts were less effective than cobalt catalysts supported on either mesoporous silica or zeolite MCM–41.

Zeolites were then found to promote the Pauson–Khand reaction *(107)*, and allowed to perform the reaction in the absence of an atmosphere of carbon monoxide, since the zeolite could previously absorb the amount of carbon monoxide required to carry out the reaction *(108,109)*.

4.3 Nazarov reaction

The Nazarov cyclization is a method to prepare cyclopentenones starting from open pentadienones, usually catalyzed by metal species *(110)* or very strong acids such as triflic acid *(111)*. In this sense, it would be expected that solid acids were not able to catalyze the reaction, since solid acids cannot arrive to such high acidicites. However, as we have seen above, the carbocation stabilization promoted by the anionic zeolite framework can override this lack of acidity to boost carbocation–mediated transformation, in this case carbo–oxonium mediated transformations *(112)*. Although not strictly a Nazarov reaction, a precedent example already showed related cyclizations

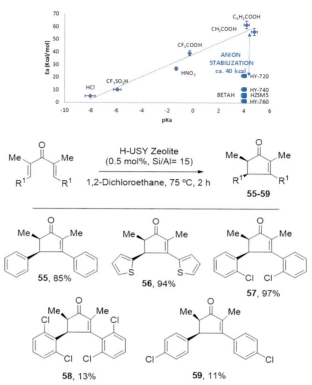

Fig. 19 Top: Activation energy (Ea) for the Nazarov cyclization to give enone **55** as a function of the acidity (pKa) of different soluble and solid acids. The Si/Al ratios for the different zeolites is: HY–720, BETA–H and ZSM–5–15, HY–740–20, and HY–760–30. Error bars account for a 5% uncertainty. Bottom: Nazarov cyclization catalyzed by H–USY zeolite under the reaction conditions indicated. GC yields. Only the *cis–cis* product is shown, since it typically accounts for >90% of the total isomeric mixture. *Adapted from Tejeda-Serrano, M.; Sanz-Navarro, S.; Blake, F.; Leyva-Pérez, A. Synth.* **2020**, *52 (14), 2031–2037.*

by employing acidic zeolites as a catalyst *(80)*. Fig. 19 shows that the activation energy for the Nazarov reaction to give cyclopentenone **55** (see reaction conditions below) is significantly decreased when diverse acid zeolites are used, including proton Y, Beta and ZSM–5 types *(113)*. Fig. 19 also shows that different cyclopentenones can be prepared with the H–USY zeolite catalyst. As it occurred with propragyl alcohols, the regents and products involve in the Nazarov reaction can unlikely penetrate into the zeolite framewrok, morevoer considering that the ZSM–5 zeolite with a narrow open pore of ~5.5 Å also catalyzes the reaction. Therefore, the outer surface of the zeolite must play a role. Indeed, previous examples of solid catalyst for this reaction required open solid catalytic surfaces, such as sulfonic acids anchored on cellulosic supports *(114)*.

5. Imination reactions

Imines are very useful and versatile intermediates in organic synthesis *(115–117)*. Conventionally, they have been obtained by condensation of amines with aldehydes or ketones *(118)*. However, new methodologies have been developed to obtain these compounds, such as the cross–coupling of alcohols with amines mediated by metals, which is a direct and simple reaction, in which aldehyde intermediates are formed and water is obtained as a byproduct, as it can be seen in Fig. 20 *(119)*. Despite being a very attractive method, finding mild reaction conditions for the selective oxidation of alcohols to aldehyde intermediates remains a challenge, since the conditions reported do not usually have tolerance to many functional groups.

There are very few examples of the use of zeolites to catalyze the preparation of imines from alcohols and amines. In 2013, a method was reported using nanosized zeolite Beta to carry out the *N*–alkylation of amines with alcohols *(120)*. However, when the scope of nanosized zeolite Beta catalyzed *N*–alkylation with various amines and alcohols was explored, it was discovered that reacting nitrobenzyl alcohols with different substituted anilines gave imines, but under very specific conditions. That is, when the reaction was carried out under an inert atmosphere, the *N*–alkylated product was formed, while if the reaction was carried out under air, the corresponding imine was obtained.

On the other hand, solid catalysts of different metals (Ni, Cr, Cu, Fe or Co) were designed by introducing the respective cations in different zeolites (13×, NaY, Na–MOR, Na–ZSM–5 and Na–Beta) by ion exchange *(121)*. These catalysts were tested to obtain imines by cross–coupling of alcohols with aromatic anilines, observing that the most active catalysts were those of Co, regardless of the type of zeolite used, obtaining the best results with the Co–13× catalyst. The reactions were carried out at 433 K and in the

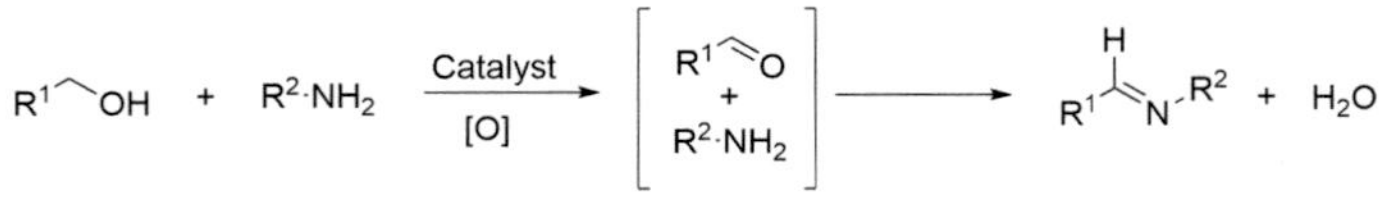

Fig. 20 Imine formation by cross–coupling of alcohols with amines.

presence of the basic promoter since, in the absence of base, the aniline conversion was low (<25%). In addition, it was found that Co–13× could be reused during 5 cycles, without leaching.

In addition, different Ru complexes have been explored to obtain imines from alcohols, but the reactions were carried out under inert atmospheres or relatively harsh conditions *(122,123)*. It should be noted that Ru complexes under an inert atmosphere can promote imine reduction by in–situ formation of hydrides during the reaction through a hydrogen–borrowing process *(124)*. A recently reported example takes a step forward, showing how the use of zeolites or MOFs to stabilize $Ru[(H_2O)_6]^{3+}$ species allows the obtaining of imines from primary alcohols under mild conditions *(125)*. It was tried to carry out the imidation of 4–vinylbenzyl alcohol and 3,4–dimethylaniline with different Ru catalysts, observing how Ru $[(H_2O)_6]\cdot3BF_4$ in solution rapidly losses catalytic activity due to the degradation of the catalyst. Using amorphous inorganic oxides as supports for this Ru species, the same was observed. On the contrary, good yields were obtained using as a support faujasite–type zeolites (NaY or KY) or a MOF. Fig. 21 shows the compatibility of the designed catalyst and the reaction conditions (slightly acid pH and low oxygen atmospheres) with a wide variety of functional groups, to give a plethora of new imines **60–80** in good yields.

6. Selective alkylation reactions

The Friedel–Crafts alkylation is a well–known reaction in zeolite catalysis, since the Brönsted acidity of the zeolite, in combination or not with confinement effects, triggers the reaction. However, the alkylation of acidic CH groups is, perhaps, a more challenging reaction for zeolite catalysts since, in contrast not only to the Friedel–Crafts alkylation but also to all the previous reactions commented above, they are usually catalyzed by bases instead of acids. Despite basic zeolites are well–known and easy to prepare, for instance by exchange of the extra–framework compensating alkaline cations form Li^+ to Cs^+ (in order to increase basicity) *(126)*, the basicity of this zeolite is not such as high to easily deprotonate typically required functional groups such as the CHs in alpha position to carbonyl groups, neither to alcohols. However, other strategies have been employed and they will be commented below.

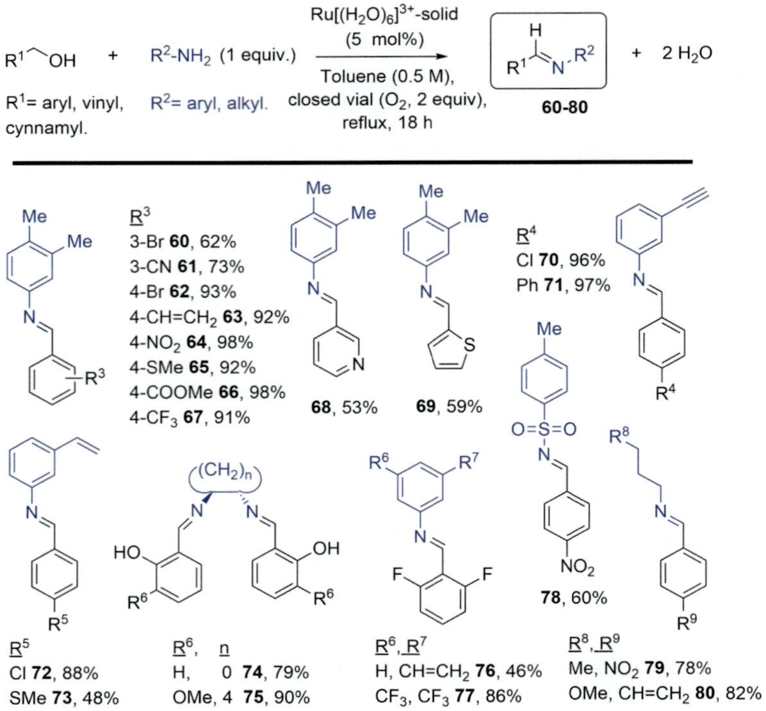

Fig. 21 Scope for the imination of primary alcohols with amines to give products **60–80**, using Ru[(H$_2$O)$_6$]$^{3+}$ species supported on NaY, KY or MOF. *Adapted from Mon, M.; Adam, R.; Ferrando-Soria, J.; Corma, A.; Armentano, D.; Pardo, E.; Leyva-Pérez, A. ACS Catal. 2018, 8 (11), 10401–10406.*

6.1 Alpha–alkylation of ketones

The alkylation of ketones is not in principle amenable to be catalyzed by basic zeolites, however, if the temnperature is sufficiently increased, this reaction can occur at some extent. One early example was reported by the group of Davies and consisted in the use of ion–exchanged CsNaX and CsNaY, and also CsOAc–impregnated CsNaX and CsNaY, to catalyze the alkylation of acetone with methanol to methyl vinyl ketone *(127)*. However, the majority of the reacted acetone formed aldol condensation products. A second strategy employed to alkylated ketones with basic zeolite catalysts was using formaldehyde instead of methanol as the alkylating agent *(128)*. In this case, acetophenone was employed as a substrate, to obtain propiophenone as a product.

Zeolites catalyze selective reactions of large organic molecules 91

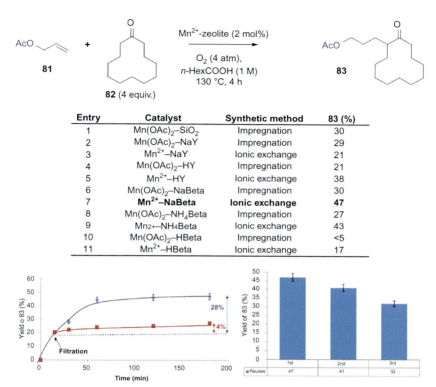

Fig. 22 Top: catalytic results for the α–alkylation radical coupling of allyl acetate **81** with cyclododecanone **82** under aerobic conditions. Bottom: Hot filtration test (left) and reuses (right) of Mn^{2+}–NaBeta catalyst during the same reaction under optimized reaction conditions. Error bars account for 5% uncertainty. *Adapted from Sanz-Navarro, S.; Garnes-Portolés, F.; López-Cruz, C.; Espinós-Ferri, E.; Corma, A.; Leyva-Pérez, A. Appl. Catal. A.* **2021**, *613, 118021.*

A different and recent strategy consisted in using Mn^{2+}–loaded zeolites to trigger a radical reaction between ketones and alkenes *(129)*. This strategy has the advantage of not generating waste products and not requiring basicity in the zeolite. Fig. 22 shows that that different zeolites exchanged with Mn^{2+}, including Na$^+$ and NH$_4^+$ containing zeolites Y and Beta, worked well as a catalyst for the coupling of allyl acetate **81** and cyclododecanone **82**, to give the alkylated product **83**. Although both the impregnated and the exchanged zeolites, except for zeolite H–Beta, gave similar yields, hot filtration tests revelead than Mn^{2+} species were only stable in the exchanged zeolites under reaction conditions, as also shown in Fig. 22. Indeed, the exchanged zeolite Mn^{2+}–NaBeta gave a 48% yield of **83**, better than any homogeneous catalyst tested. These results strongly suggest that the zeolite framework can isolate Mn cations by ion exchange, stabilizing Mn^{2+} to efficiently catalyze the radical reaction *(130)*. These results connect with the fact that zeolites are active matrixes for radical reactions of ketones and aldol–type transformations *(131)*.

6.2 Alkylation of arenes and alcohols

The Friedel–Crafts alkylation of arenes is perhaps one of the best studies processes in zeolites, with application in fine chemsitry. This reaction is industrially carried out with $AlCl_3$, thus it is not surprising that, from very early times, zeolites were studied as potential alternatives to the corrosive soluble catalysts. Alcohols are used here as alkylating agents. However, the alkylation of alcohols is a less studied process (if we do not consider the Friedel–Crafts reaction as a method of alkylation, since arenes and not alkyl groups are incorporated in the alcohol), but some advances have been made in the last years. The alkylation of amines is also a fundamental organic reaction to further funcionalize relevant molecules in fine chemistry and pharma, however, this topic has been recently reviewed and will not be treated here *(132)*.

The search of accesible zeolites for bulky molecules is one of the most active fields of research in zeolite catalysis, and key to apply zeolites in fine cheisitry. It must be noticed here that the alkylation reactions with zeolites are perhaps where the search for these accesible sites has been more pronounced, since the alkylation of arenes and alcohols employ relatively bulky reagents which require good diffusivity along the pores. The last advances in this regard are commented below.

6.2.1 Friedel–Crafts alkylation

Advances in the last 3 years for this reaction are commented in this section. The first of these studies already intoduces the concept of not–limiting substrate size, since the new zeolite system can make diffuse and react big reagents that otherwise will not penetrate through the pores of conventional zeolites *(133)*. For that, nanocrystallyne ZSM–5 and Beta zeolites with trimodal porosity were synthesized by using starch coated silica nanospheres as silica precursors. The final interparticle porosity was 50–200 nm and, besides, intercrystal mesopores of 2–50 nm were generated. The new zeolites showed excellent acidity and catalytic activity in the liquid phase Friedel–Crafts alkylation followed by ring closure reaction for the synthesis of vitamin E. Comparison with the corresponding conventional zeolites showed excellent recyclability for the former.

A good accesibility of bulky reactants into the zeolite was also achieved by the group of Jones, employing two–dimensional (2D) zeolites *(134)*, in this work, 2D MFI zeolite nanosheets, composed of multilamellar stacks of MFI nanosheets, were employed as catalysts for the liquid phase Friedel–Crafts alkylation of mesitylene with benzyl alcohol, a reaction which cannot occur in the micropores of typical zeolites. However, the self–etherification of benzyl alcohol also occurred as a secondary reaction. Both processes are catalyzed by the Bronsted acid sites of the zeolite. It was found that the etherification reaction was the main contributor to the decrease of the turn-over frequency (TOF) when decreasing the Si/Al ratio of the zeolite. Remarkable, when the same reaction was carried out in the presence of a bulky poison such 2,6 di–*tert*–butylpyridine (DTBP), only the etherification reaction of benzyl alcohols takes place, since DTBP deactivates the external surface acid sites.

In order to further study the above results, the group of Tsapatsis used superheated steam to treat the hierarchical MFI zeolites, to alter the nanosheet morphololy and regulate the external surface catalytic activity, while preserving micro– and mesoporosity *(135)*. Characterization mea-surements showed a nearly invariant structure for the two dimensional nanosheet MFI zeolites before and after the superheated steam treatment. However, a pronounced reaction rate decrease for benzyl alcohol alkylation occurred. Transmission electron microscopy images revealed nanosheet coarsening with the corresponding external (mesoporous) surface structure and catalytic activity modification. These results illustrate the differences between the catalytic sites inside and outside the zeolite channels after cehmcail treatments, and their potential selective modulation.

The self–etherification of the alkylating agent alcohol is a recurrent problem in the Friedel–Crafts alkylation, as we have seen in the above example, thus the selectivity of the catalyst is key. The incorporation of Cu species into hierarchical zeolites by a simple postdesilication–recrystallization strategy has allowed to perform the reaction above (liquid alkylation reaction of mesitylene with benzyl alcohol) with extraordinary selectivity, particularly compared to the bare zeolite *(136)*. Studies with dif-ferent amounts of Cu showed that finely tuning the acidity of the zeolite, with the right balance between the redox properties of Cu and the acidity of the zeolite, allows to get an optimum catalytic performance for this reaction.

Following with the search of selective zeolites able to incorporate and make react the bulky molecules participating in the Friedel–Crafts alkyl-ation, particualrly thinking in lignocellulosic arenes and alcohols, mesoporous and microporous H–Beta zeolites were prepared and used as catalysts for the reaction of different arenes (benzene, toluene, p–xylene and mesitylene) with benzyl alcohol *(137)*. The number of catalytically active Bronsted acid sites were determined by ^{1}H magic–angle spinning (MAS) NMR experiments, and the results showed that introducing meso-pores into the H–Beta zeolite structure significantly increased the catalytic performance for bulky arenes such as mesytilene, but not for small arenes such as benzene. This study is another example of how the required substrate size control of the reaction points out which zeolite catalyst to use.

6.2.2 Alcohol alkylation

The alkylation of alcohols is a fundamental organic reaction in fine chemsitry and pharma, traditionally performed with strong bases. Nevertheless, recent examples demonstrate that zeolites can be employed as catalysts for this reaction. A remarkable example by the group of Lercher *(138)* have shown that the alkylation of phenol with cyclohexanol in the liquid phase can be perfomed with zeolites in apolar solvents, due to the formation of carbenium ions in the Brönsted acid sites. When the zeolite pores are filled with water, the formation of the electrophilic carbenium decreases. The alkylation of phenolics is of great importance in synthetic chemistry due to the valoriza-tion of lignocellulosic–biomass–derived phenols, thus it is not surprising that the alkylation reaction of these phenols deserves further studies. Another recent work by the same group shows that zeolites HBEA and HY are oper-ative as catalysts for the efficient alkylation of phenols due to the presence of moderately strong Brönsted acid sites in their structures, while very strong acid sites seemed to be responsible for catalyst deactivation *(139)*. Besides, solids without molecularly–sized pore constraints showed lower Brönsted acid sites and, consequently, low turnover frequencies.

The alkylation of benzyl alcohols with *tert*–butyl acetate in the presence of zeolite catalysts has also been recently studied *(113)*. The use of *tert*–butyl acetate as an alkylating agent circumvents not only the use of classical alkylating agents under basic conditions but also the use of other alkylating alcohols under acid conditions, with the recurrent self–etherification reac-tion. Here, the acetate is able to generate carbo–oxonium species under acid conditions, which eventually release *tert*–butyl cations to the reaction medium, rapidly cathched by benzyl alcohol. Since carbo–oxonium species must be generated in–situ, it was envisioned the use of zeolites as generator agents, which have proven succesful in generating and stabilizing such spe-cies during the Nazarov reaction. Fig. 23 shows that, as it occurred for the intramolecular hydroalkoxylation and hydrothioalkoxylation of propargyl

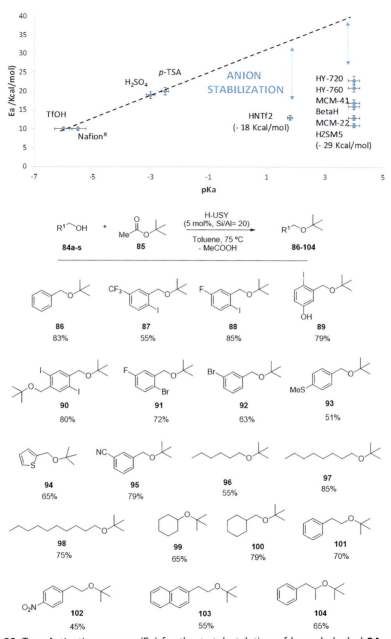

Fig. 23 Top: Activation energy (Ea) for the *tert*–butylation of benzyl alcohol **84a** with *tert*–butyl acetate **85** to give the ether product **86**, as a function of the acidity (pKa) of different soluble and solid acids. Error bars account for a 5% uncertainty. Bottom: *Tert*–butylation of benzyl, alkyl and homoallyl alcohols **84a–s** with *tert*–butyl acetate **85** catalyzed by H–USY zeolite (Si/Al ratio ∼ 20), to give *tert*–butyl ethers **86–104**. GC yields. *Adapted from Tejeda-Serrano, M.; Sanz-Navarro, S.; Blake, F.; Leyva-Pérez, A. Synth.* **2020**, *52 (14), 2031–2037.*

(thio)amides, and also in the Nazarov reaction, the activation energy of the reaction significantly decreases for the zeolite catalyst when represented vs the expected pKa of their acid sites, compared with the linear behavior of the soluble sulfonic acids. A variety of *tert*–butyl ethers **86–104** could be prepared by using zeolite HY as a catalyst for this reaction (see below).

7. Conclusions and outlook

Zeolites are efficient catalysts for advanced organic reactions, of application in fine chemical synthesis and pharma. Despite the inherent acid sites in protic zeolites are the most used catalytic sites, a plethora of synthetic strategies have been employed to get new catalytic sites either in the pores, cavities, outer surface and even interparticle space of the zeolite. This synthetic strategies include: (i) the incorporation of metal sites either by cationic exchange, direct impregnation or covalent immobilization (grafting), (ii) the tuning of the zeolite framework charge with the charge–balancing exchangeable cations, thus allowing to have from very acid to moderately basic zeolites, (iii) the generation of hierarchical zeolites with different porosities, to regulate the molecular traffic across the zeolite, (iv) the use of nanocrystalline, 2D and assembled zeolites to increase the accessibility of bulk reactants, beyond using the outer surface and interparticle space of conventional zeolites. All these techniques sum to the native shape selectivity of zeolites as a function of its structure and the archetypical chemical modification of the zeolite framework with different atoms (doping, generation of extra–framework species, degree of hydration, …).

The metathesis reaction has been accomplished with zeolite catalysts in both the classical alkene–alkene and the newer alkene–carbonyl versions. While the former seems to require a metal site, the later can be accomplished with pure acid zeolitic materials, preferentially with the cheap and widely available montmorillonite solid material. Different hydroaddition reactions to unsaturated carbon–carbon bonds, including the formation of new C—B, C—Si, C—O, C—S, C—N and C—P bonds, are catalyzed by zeolites, typically with metals incorporated in its structure, however, the Brönsted sites of conventional acid zeolites can be used in some cases. The spatial constraints of the zeolite pores can be employed not only for substrate shape selectivity, but also to generate selectively metal single atoms and clusters within the zeolite cavities and channels to regioselectively catalyze certain hydroaddition reactions, i.e., the Markovnikov hydrosilylation of alkynes. Electrocyclic reactions can also be done within the zeolite pores, for instance the Diels–Alder reaction has been deeply studied. However, other electrocyclic reactions can be catalyzed in the outer surface or even in the interparticle space of the zeolite, such as the Nazarov reaction. The imination reaction is another process that can be catalyzed by metal–substituted zeolites, for instance with Ru, promoting hydrogen–borrowing processes inside the zeolite. Finally, we have seen here that the alkylation in alpha positions to carbonyl compounds, and also of arenes (Friedel–Crafts reaction) and alcohols, proceeds in the presence of basic and acid zeolites, with particular emphasis in the generation of widely accessible zeolites for bulky reactants. The stabilization of carbenium and carbo–oxonium species by the anionic framework of the zeolitic framework plays a key role in those reactions where these species are intermediates. This issue is especially relevant when translating reactions typically carried out with very strong soluble acids such as sulfonic, sulfuric and phosphoric acid, not only in practical terms but also considering mechanistic richness during the organic reaction, since the zeolite stabilization can open new reaction pathways. Soluble acids are intensively employed in the chemical industry due to their low price but the present clear inherent risks in terms of handing, storage and waste; in contrast, zeolites are more expensive but they can be reused, implemented in continuous processes, easily managed and stored, thus providing a real alternative to those soluble corrosive acids.

If one considers all these synthetic strategies with the inherent robustness, low toxicity, relatively cheap price and recyclability of the zeolite, and also the possibility of easily designing in–flow processes with these solid catalysts, which have been already demonstrated in the petrochemical industry, the possibilities are endless. Therefore, it can be foreseen that zeolites will increase their use as catalysts in advanced organic synthesis in the last years.

Acknowledgments

M. M. thanks to the Ministerio de Innovación y Ciencia (Ayudas Juan de la Cierva-Formación 2019, Project number FJC2019-040523-I).

References

1. Van Bekkum, H.; Kouwenhoven, H. W. Zeolites and fine chemicals. In *Heterogeneous Catalysis and Fine Chemicals*; Guisnet, M., Barrault, J., Bouchoule, C., Duprez, D., Montassier, C., Pérot, G., Eds.; Elsevier Science: Amsterdam, 1988; pp. 45–59.
2. Dartt, C. B.; Davis, M. E. *Catal. Today* **1994**, *19*, 151–186.
3. Davis, M. E. *Microporous Mesoporous Mater.* **1998**, *21*, 173–182.
4. Sheldon, R. A. *Stud. Surf. Sci. Catal.* **1991**, *59*, 33–54.
5. Hölderich, W. F.; van Bekkum, H. *Stud. Surf. Sci. Catal.* **1991**, *58*, 631–726.
6. Espeel, P. H.; Janssens, B.; Jacobs, P. A. *J. Org. Chem.* **1993**, *58*, 7688–7693.

7. Espeel, P. H. J.; Vercruysse, K. A.; Debaerdemaker, M.; Jacobs, P. A. *Stud. Surf. Sci. Catal.* **1994**, *84*, 1457–1464.
8. Mirth, G.; Cejka, J.; Lercher, J. A. *J. Catal.* **1993**, *139*, 24–33.
9. Armengol, E.; Cano, M. L.; Corma, A.; García, H.; Navarro, M. T. *J. Chem. Soc., Chem. Commun.* **1995**, 519–520.
10. Corma, A.; Sastre, E. *J. Catal.* **1991**, *129*, 177–185.
11. Corma, A.; Iborra, S. Zeolites and related materials in Knoevenagel condensations and Michael additions. In *Fine Chemicals Through Heterogeneous Catalysis*; Sheldon, R. A., Van Bekkum, H., Eds.; Wiley-VCH: Weinheim, 2001; pp. 309–326.
12. Weigert, F. J. *J. Org. Chem.* **1987**, *52*, 3296–3298.
13. Corma, A.; Renz, M. *Chem. Commun.* **2004**, 550–551.
14. Oudejans, J. C.; van Bekkum, H. *J. Mol. Catal.* **1981**, *12*, 149.
15. Arias, K. S.; Climent, M. J.; Corma, A.; Iborra, S. *ACS Sustainable Chem. Eng.* **2016**, *4*, 6152–6159.
16. Tao, L.; Wang, Z.-J.; Yan, T.-H.; Liu, Y.-M.; He, H.-Y.; Cao, Y. *ACS Catal.* **2017**, *7*, 959–964.
17. Subbotina, E.; Velty, A.; Samec, J. S. M.; Corma, A. *ChemSusChem* **2020**, *13*, 4528–4536.
18. Banks, R. L.; Bailey, G. C. *Ind. Eng. Chem. Prod. Res. Dev.* **1964**, *3*, 170–173.
19. Hérisson, J. L.; Chauvin, Y. *Makromol. Chem.* **1971**, *141*, 161–176.
20. Singh, O. M. *J. Sci. Ind. Res.* **2006**, *65*, 957–965.
21. Mol, J. C. *J. Mol. Catal. A: Chem.* **2004**, *213*, 39–45.
22. Grubbs, R. H.; O'Leary, D. L. Handbook of metathesis. In *Applications in Organic Synthesis*; *Vol. 2*; Wiley–VCH: Weinheim, 2015.
23. Chauvin, Y. *Angew. Chem. Int. Ed.* **2006**, *45*, 3741–3747.
24. Copéret, C.; Allouche, F.; Chan, K. W.; Conley, M. P.; Delley, M. F.; Fedorov, A.; Moroz, I. B.; Mougel, V.; Pucino, M.; Searles, K.; Yamamoto, K.; Zhizhko, P. A. *Angew. Chem., Int. Ed.* **2018**, *57*, 6398–6440.
25. Lwin, S.; Wachs, I. E. *ACS Catal.* **2014**, *4*, 2505–2520.
26. Mougel, V.; Chan, K.-W.; Siddiqi, G.; Kawakita, K.; Nagae, H.; Tsurugi, H.; Mashima, K.; Safonova, O.; Copéret, C. *ACS Cent. Sci.* **2016**, *2*, 569–576.
27. Amigues, P.; Chauvin, Y.; Commereuc, D.; Hong, C. T.; Lai, C. C.; Liu, Y. H. *J. Mol. Catal.* **1991**, *65*, 39–50.
28. Spronk, R.; Andreini, A.; Mol, J. C. *J. Mol. Catal.* **1991**, *65*, 219–235.
29. Hamdan, H.; Ramli, Z. *Stud. Surf. Sci. Catal.* **1997**, *105*, 957–964.
30. Bein, T.; Huber, C.; Moller, K.; Wu, C.-G.; Xu, L. *Chem. Mater.* **1997**, *9*, 2252–2254.
31. Consoli, D. F.; Zhang, S.; Shaikh, S.; Román-Leshkov, Y. *Org. Process Res. Dev.* **2018**, *22*, 1683–1686.
32. Zhao, P.; Ye, L.; Sun, Z.; Lo, B. T. W.; Woodcock, H.; Huang, C.; Tang, C.; Kirkland, A. I.; Mei, D.; Edman Tsang, S. C. *J. Am. Chem. Soc.* **2018**, *140*, 6661–6667.
33. Zhao, P.; Ye, L.; Li, G.; Huang, C.; Wu, S.; Ho, P.-L.; Wang, H.; Yoskamtorn, T.; Sheptyakov, D.; Cibin, G.; Kirkland, A. I.; Tang, C. C.; Zheng, A.; Xue, W.; Mei, D.; Suriye, K.; Edman Tsang, S. C. *ACS Catal.* **2021**, *11*, 3530–3540.
34. Blay, V.; Peeled, E.; Miravalles, R.; Perea, L.; Perea, A. *Catal. Rev.* **2018**, *60*, 278–335.
35. Pastva, J.; Skowerski, K.; Czarnocki, S. J.; Žilková, N.; Čejka, J.; Bastl, Z.; Balcar, H. *ACS Catal.* **2014**, *4*, 3227–3236.
36. Balcar, H.; Žilková, N.; Kubů, M.; Mazur, M.; Bastl, Z.; Čejka, J. *Beilstein J. Org. Chem.* **2015**, *11*, 2087–2096.
37. Díaz, U.; Corma, A. *Dalton Trans.* **2014**, *43*, 10292–10316.
38. Roth, W. J.; Nachtigall, P.; Morris, R. E.; Čejka, J. *Chem. Rev.* **2014**, *114*, 4807–4837.
39. Ludwig, J. R.; Schindler, C. S. *Synlett* **2017**, *28*, 1501–1509.

40. Ludwig, J. R.; Watson, R. B.; Nasrallah, D. J.; Gianino, J. B.; Zimmerman, P. M.; Wiscons, R. A.; Schindler, C. S. *Science* **2018**, *361*, 1363–1369.
41. van Schaik, H.-P.; Vijn, R.-J.; Bickelhaupt, F. *Angew. Chem. Int. Ed. Engl.* **1994**, *33*, 1611–1612.
42. Lewis, J. D.; Van de Vyver, S.; Román-Leshkov, Y. *Angew. Chem., Int. Ed.* **2015**, *54*, 9835–9838.
43. Rivero-Crespo, M. A.; Tejeda-Serrano, M.; Pérez-Sánchez, H.; Cerón-Carrasco, J. P.; Leyva-Pérez, A. *Angew. Chem., Int. Ed.* **2020**, *59*, 3846–3849.
44. Alonso, F.; Beletskaya, I. P.; Yus, M. *Chem. Rev.* **2004**, *104*, 3079–3159.
45. Corma, A.; Leyva-Pérez, A.; Sabater, M. J. *Chem. Rev.* **2011**, *111*, 1657–1712.
46. Leyva-Pérez, A.; Corma, A. *Angew. Chem., Int. Ed.* **2012**, *51*, 614–635.
47. Brown, H. C.; Chen, G. M.; Jennings, M. V.; Ramachandran, P. V. *Angew. Chem., Int. Ed.* **1999**, *38*, 2052–2054.
48. Crudden, C. M.; Edwards, D. *Eur. J. Org. Chem.* **2003**, 4695–4712.
49. Carmona, A.; Corma, A.; Iglesias, M.; San Jose, A.; Sanchez, F. *J. Organometal. Chem.* **1995**, *492*, 11–21.
50. Segarra, A. M.; Guerrero, R.; Claver, C.; Fernandez, E. *Chem. Eur. J.* **2003**, *9*, 191–200.
51. Jimenez, O.; Mueller, T. E.; Schwieger, W.; Lercher, J. A. *Stud. Surf. Sci. Catal.* **2004**, *154C*, 2788–2794.
52. Penzien, J.; Muller, T. E.; Lercher, J. A. *Chem. Commun.* **2000**, *18*, 1753–1754.
53. Penzien, J.; Muller, T. E.; Lercher, J. A. *Micropor. Mesopor. Mater.* **2001**, *48*, 285–291.
54. Penzien, J.; Haessner, C.; Jentys, A.; Kohler, K.; Muller, T. E.; Lercher, J. A. *J. Catal.* **2004**, *221*, 302–312.
55. Shanbhag, G. V.; Joseph, T.; Halligudi, S. B. *J. Catal.* **2007**, *250*, 274–282.
56. Horniakova, J.; Komura, K.; Osaki, H.; Kubota, Y.; Sugi, Y. *Catal. Lett.* **2005**, *102*, 191–196.
57. Jimenez, O.; Mueller, T. E.; Schwieger, W.; Lercher, J. A. *J. Catal.* **2006**, *239*, 42–50.
58. Kore, R.; Satpati, B.; Srivastava, R. *Appl. Catal. A* **2014**, *477*, 8–17.
59. Ciobanu, M.; Tirsoaga, A.; Amoros, P.; Beltran, D.; Coman, S. M.; Parvulescu, V. I. *Appl. Catal. A* **2014**, *474*, 230–235.
60. Komura, K.; Hongo, R.; Tsutsui, J.; Sugi, Y. *Catal. Lett.* **2009**, *128*, 203–209.
61. Sudheesh, N.; Shukla, R. S. *Appl. Catal. A* **2014**, *473*, 116–124.
62. Kore, R.; Srivastava, R.; Satpati, B. *ACS Catal.* **2013**, *3*, 2891–2904.
63. Kore, R.; Srivastava, R.; Satpati, B. *Chem. Eur. J.* **2014**, *20*, 11511–11521.
64. Penzien, J.; Su, R. Q.; Muller, T. E. *J. Mol. Catal. A* **2002**, *182–183*, 489–498.
65. Corma, A.; Gonzalez-Arellano, C.; Iglesias, M.; Navarro, M. T.; Sanchez, F. *Chem. Commun.* **2008**, *46*, 6218–6220.
66. Sazama, P.; Wichterlova, B.; Sklenak, S.; Parvulescu, V. I.; Candu, N.; Sadovska, G.; Dedecek, J.; Klein, P.; Pashkova, V.; Stastny, P. *J. Catal.* **2014**, *318*, 22–33.
67. Rubio-Marqués, P.; Rivero-Crespo, M. A.; Leyva-Pérez, A.; Corma, A. *J. Am. Chem. Soc.* **2015**, *137*, 11832–118379.
68. Wang, X.; Wang, H.; Zhao, K.; Li, T.; Liu, S.; Yuan, H.; Shi, F. *J. Catal.* **2021**, *394*, 18–29.
69. Leyva, A.; Corma, A. *J. Org. Chem.* **2009**, *74*, 2067–2074.
70. Corma, A.; Ruiz, V. R.; Leyva-Pérez, A.; Sabater, M. J. *Adv. Synth. Catal.* **2010**, *352*, 1701–1710.
71. Finiels, A.; Geneste, P.; Marichez, F.; Moreau, P. *Catal. Lett.* **1989**, *2*, 181–184.
72. Sartori, G.; Pastorio, A.; Magi, R.; Bigi, F. *Tetrahedron* **1996**, *52*, 8287–8296.
73. Mameda, N.; Peraka, S.; Marri, M. R.; Kodumuri, S.; Chevella, D.; Gutta, N.; Nama, N. *Appl. Catal. A* **2015**, *505*, 213–216.

74. Mameda, N.; Peraka, S.; Kodumuri, S.; Chevella, D.; Banothu, R.; Amrutham, V.; Nama, N. *RSC Adv.* **2016**, *6*, 58137–58141.
75. Sultana Poly, S.; Hakim Siddiki, S. M. A.; Touchy, A. S.; Yasumura, S.; Toyao, T.; Maeno, Z.; Shimizu, K.–i. *J. Catal.* **2018**, *368*, 145–154.
76. Xu, S.; Yun, Z.; Feng, Y.; Tang, T.; Fang, Z.; Tang, T. *RSC Adv.* **2016**, *6*, 69822–69827.
77. Pérez-Mayoral, E.; Matos, I.; Nachtigall, P.; Polozij, M.; Fonseca, I.; Vitvarova-Prochazkova, D.; Cejka, J. *ChemSusChem* **2013**, *6*, 1021–1030.
78. Radhakrishnan, S.; Thoelen, G.; Franken, J.; Degreve, J.; Kirschhock, C. E. A.; Martens, J. A. *ChemCatChem* **2013**, *5*, 576–581.
79. Zamanian, S.; Kharat, A. N. *Austral. J. Chem.* **2015**, *68*, 981–986.
80. Cabrero-Antonino, J. R.; Leyva-Pérez, A.; Corma, A. *Angew. Chem., Int. Ed.* **2015**, *54*, 5658–5661.
81. Delacroix, O.; Gaumont, A. C. *Curr. Org. Chem.* **2005**, *9*, 1851–1882.
82. Leyva-Pérez, A.; Vidal-Moya, J. A.; Cabrero-Antonino, J. R.; Al-Deyab, S. S.; Al-Resayes, S. I.; Corma, A. *J. Organometal. Chem.* **2011**, *696*, 362–367.
83. Yao, F.; Peng, J.; Hao, W.; Cai, M. *Catal. Lett.* **2012**, *142*, 803–808.
84. Huang, Y.; Hao, W.; Ding, G.; Cai, M.-Z. *J. Organometal. Chem.* **2012**, *715*, 141–146.
85. Obligacion, J. V.; Chirik, P. J. *Nat. Rev. Chem.* **2018**, *2*, 15–34.
86. Pappas, I.; Treacy, S.; Chirik, P. J. *ACS Catal.* **2016**, *6*, 4105–4109.
87. Pagliaro, M.; Ciriminna, R.; Pandarus, V.; Béland, F. *Eur. J. Org. Chem.* **2013**, 6227–6235.
88. Fiedorow, R.; Wawrzynczak, A. Catalysts for Hydrosilylation. In *Heterogeneous Systems*; Marciniec, B., Ed.; Education in Advanced Chemistry, 10; Wydawnictwo Poznanskie: Poznan, Poland, 2006; pp. 327–344.
89. Cui, X.; Junge, K.; Dai, X.; Kreyenschulte, C.; Pohl, M.-M.; Wohlrab, S.; Shi, F.; Bruckner, A.; Beller, M. *ACS Cent. Sci.* **2017**, *3*, 580–585.
90. Vivero-Escota, J. L.; Wo, L. K. *Chemtracts* **2006**, *19*, 358–366.
91. Corma, A.; González-Arellano, C.; Iglesias, M.; Sánchez, F. *Angew. Chem., Int. Ed.* **2007**, *46*, 7820–7822.
92. Do Van, D.; Hosokawa, T.; Saito, M.; Horiuchi, Y.; Matsuoka, M. *Appl. Catal. A: Gen.* **2015**, *503*, 203–208.
93. Garcés, K.; Fernández-Alvarez, F. J.; García-Orduña, P.; Lahoz, F. J.; Pérez-Torrente, J. J.; Oro, L. A. *ChemCatChem* **2015**, *7*, 2501–2507.
94. Hu, R.; Hao, W.; Cai, M. *Chin. J. Chem.* **2011**, *29*, 1629–1634.
95. Corma, A.; de Dios, M. I.; Iglesias, M.; Sánchez, F. *Stud. Surf. Sci. Catal.* **1997**, *108*, 501–507.
96. Rivero-Crespo, M.; Oliver-Meseguer, J.; Kapłońska, K.; Kuśtrowski, P.; Pardo, E.; Cerón-Carrasco, J. P.; Leyva-Pérez, A. *Chem. Sci.* **2020**, *11*, 8113–8124.
97. Kumar, P.; Kumar, P. R.; Hegde, V. R. *Synlett* **1999**, *12*, 1921–1922.
98. Kanagasabapathy, S.; Sudalai, A.; Benicewicz, B. C. *Tetrahedron Lett.* **2001**, *42*, 3791–3794.
99. Ogawa, A.; Ikeda, T.; Kimura, K.; Hirao, T. *J. Am. Chem. Soc.* **1999**, *121*, 5108–5114.
100. Shoai, S.; Bichler, P.; Kang, B.; Buckley, H.; Love, J. A. *Organometallics* **2007**, *26*, 5778–5781.
101. Zhao, H.; Peng, J.; Cai, M. *Catal. Lett.* **2012**, *142*, 138–142.
102. Corma, A.; Gonzalez-Arellano, C.; Iglesias, M.; Sanchez, F. *Appl. Catal. A* **2010**, *375*, 49–54.
103. Ranu, B. C.; Chattopadhyay, K. Eco–Friendly Synthesis of Fine Chemicals. *RSC Green Chemistry Series*, Vol. 3; 2009; pp. 186–219.
104. Van de Vyver, S.; Roman-Leshkov, Y. *Angew. Chem., Int. Ed.* **2015**, *54*, 12554–12561.

105. Settle, A. E.; Berstis, L.; Rorrer, N. A.; Roman-Leshkov, Y.; Beckham, G. T.; Richards, R. M.; Vardon, D. R. *Green Chem.* **2017**, *19*, 3468–3492.
106. Kim, S.-W.; Son, S. U.; Lee, S. I.; Hyeon, T.; Chung, Y. K. *J. Am. Chem. Soc.* **2000**, *122*, 1550–1551.
107. Perez-Serrano, L.; Blanco-Urgoiti, J.; Casarrubios, L.; Dominguez, G.; Perez-Castells, J. *J. Org. Chem.* **2000**, *65*, 3513–3519.
108. Blanco-Urgoiti, J.; Casarrubios, L.; Dominguez, G.; Perez-Castells, J. *Tetrahedron Lett.* **2002**, *43*, 5763–5765.
109. Blanco-Urgoiti, J.; Dominguez, G.; Perez-Castells, J. In *Catalysts for Fine Chemical Synthesis*; Roberts, S. M., Ed.; Vol. *3*; **2004**; pp. 182–185.
110. He, W.; Sun, X.; Frontier, A. J. *J. Am. Chem. Soc.* **2003**, *125*, 14278–14279.
111. Vaidya, T.; Eisenberg, R.; Frontier, A. J. *ChemCatChem* **2011**, *3*, 1531–1548.
112. Lloret, V.; Rivero-Crespo, M. A.; Vidal-Moya, J. A.; Wild, S.; Domenech-Carbo, A.; Heller, B. S. J.; Shin, S.; Steinrueck, H.-P.; Maier, F.; Hauke, F.; Varela, M.; Hirsch, A.; Leyva-Pérez, A.; Abellan, G. N. *Nat. Commun.* **2019**, *10*, 509.
113. Tejeda-Serrano, M.; Sanz-Navarro, S.; Blake, F.; Leyva-Pérez, A. *Synthesis* **2020**, *52*, 2031–2037.
114. Daneshfar, Z.; Rostami, A. *RSC Adv.* **2015**, *5*, 104695–104707.
115. Kobayashi, S.; Mori, Y.; Fossey, J. S.; Salter, M. M. *Chem. Rev.* **2011**, *111*, 2626–2704.
116. Yao, S.; Saaby, S.; Hazell, R. G.; Jørgensen, K. A. *Chem.-Eur. J.* **2000**, *6*, 2435–2448.
117. He, R.; Jin, X.; Chen, H.; Huang, Z.-T.; Zheng, Q.-Y.; Wang, C. *J. Am. Chem. Soc.* **2014**, *136*, 6558–6561.
118. Reeves, J. T.; Visco, M. D.; Marsini, M. A.; Grinberg, N.; Busacca, C. A.; Mattson, A. E.; Senanayake, C. H. *Org. Lett.* **2015**, *17*, 2442–2445.
119. Chen, B.; Wang, L.; Gao, S. *ACS Catal.* **2015**, *5*, 5851–5876.
120. Reddy, M. M.; Kumar, M. A.; Swamy, P.; Naresh, M.; Srujana, K.; Satyanarayana, L.; Venugopal, A.; Narender, N. *Green Chem.* **2013**, *15*, 3474–3483.
121. Sun, Y.-W.; Lu, X.-H.; Wei, X.-L.; Zhou, D.; Xia, Q.-H. *Catal. Commun.* **2014**, *43*, 213–217.
122. Maggi, A.; Madsen, R. *Organometallics* **2012**, *31*, 451–455.
123. Balaraman, E.; Srimani, D.; Diskin-Posner, Y.; Milstein, D. *Catal. Lett.* **2015**, *145*, 139–144.
124. Yang, Q.; Wang, Q.; Yu, Z. *Chem. Soc. Rev.* **2015**, *44*, 2305–2329.
125. Mon, M.; Adam, R.; Ferrando-Soria, J.; Corma, A.; Armentano, D.; Pardo, E.; Leyva-Pérez, A. *ACS Catal.* **2018**, *8*, 10401–10406.
126. Garnes-Portolés, F.; Greco, R.; Oliver-Meseguer, J.; Castellanos-Soriano, J.; Consuelo Jiménez, M.; López-Haro, M.; Hernández-Garrido, J. C.; Boronat, M.; Pérez-Ruiz, R.; Leyva-Pérez, A. *Nat. Catal.* **2021**, *4*, 293–303.
127. Hathaway, P. E.; Davis, M. E. *J. Catal.* **1989**, *119*, 497–507.
128. Kulkarni, S. J.; Madhavi, G.; Rao, A. R.; Mohan, K. V. V. K. *Catal. Commun.* **2007**, *9*, 532–538.
129. Sanz-Navarro, S.; Garnes-Portolés, F.; López-Cruz, C.; Espinós-Ferri, E.; Corma, A.; Leyva-Pérez, A. *Appl. Catal. A.* **2021**, *613*, 118021.
130. Gerber, I. C.; Serp, P. *Chem. Rev.* **2020**, *120*, 1250–1349.
131. Rodriguez-Fernandez, A.; Di Iorio, J.; Paris, C.; Boronat, M.; Corma, A.; Roman-Leshkov, Y.; Moliner, M. *Chem. Sci.* **2020**, *11*, 10225–10235.
132. Bayguzina, A. R.; Khusnutdinov, R. I. *Russ. J. Gen. Chem.* **2021**, *91*, 305–347.
133. Rani, P.; Srivastava, R. *ACS Sustain. Chem. Engineer.* **2019**, *7*, 9822–9833.
134. Korde, A.; Min, B.; Almas, Q.; Chiang, Y.; Nair, S.; Jones, C. W. *ChemCatChem* **2019**, *11*, 4548–4557.

135. Guefrachi, Y.; Sharma, G.; Xu, D.; Kumar, G.; Vinter, K. P.; Abdelrahman, O. A.; Li, X.; Alhassan, S.; Dauenhauer, P. J.; Navrotsky, A.; Zhang, W.; Tsapatsis, M. *Angew. Chem., Int. Ed.* **2020**, *59*, 9579–9585.
136. Huang, J.; Liu, B.; Liao, Z.; Chen, H.; Yan, K. *Ind. Engineer. Chem. Res.* **2019**, *58*, 16636–16644.
137. Zeng, X.; Wang, Z.; Ding, J.; Wang, L.; Jiang, Y.; Stampfl, C.; Hunger, M.; Huang, J. *J. Catal.* **2019**, *380*, 9–20.
138. Liu, Y.; Barath, E.; Shi, H.; Hu, J.; Camaioni, D. M.; Lercher, J. A. *Nat. Catal.* **2018**, *1*, 141–147.
139. Liu, Y.; Cheng, G.; Barath, E.; Shi, H.; Lercher, J. A. *Appl. Catal. B* **2021**, *281*, 119424.

About the authors

Dr Marta Mon is an associated postdocoral researcher at the Institute of Chemical Technology CSIC–UPV, València. She holds a Ph.D. in Chemistry (awarded with the best thesis prize by the University of València), with expertise on materials science and coordination chemistry, particularly in the rational design of metal-organic frameworks (MOFs), and its use in diverse applications of high technological and/or environmental interest, including catalysis. Currently she is working on the preparation and evaluation of solid catalysts for organic synthesis.

Dr Antonio Leyva–Pérez is Tenured Scientist at Institute of Chemical Technology CSIC–UPV, Valencia. He holds a Ph.D in Chemistry (best thesis prize by the UPV), with expertise on materials science and organic synthesis, specifically in nanomaterials and catalyst for organic reactions. He leads the group "Catalysis for Sustainable Organic Reactions" and has published near 100 peer-reviewed scientific studies, and is co–author of 13 patents.

CHAPTER THREE

Metal-π-allyl mediated asymmetric cycloaddition reactions

Pol de la Cruz-Sánchez and Oscar Pàmies*

Universitat Rovira i Virgili, Departament de Química Física i Inorgànica, C/Marcel·lí Domingo, Tarragona, Spain
*Corresponding author: e-mail address: oscar.pamies@urv.cat

Contents

1. Introduction		104
2. Cycloaddition reactions for the formation of O-heterocycles		105
	2.1 Tetrahydrofurans	105
	2.2 Dihydrofurans	115
	2.3 Tetrahydropyrans	117
	2.4 Dioxanes	118
	2.5 Lactones	118
	2.6 Oxepines	123
	2.7 Dioxonines	124
3. Cycloaddition reactions for the formation of N-heterocycles		124
	3.1 Pyrrolidines	124
	3.2 Indoles	130
	3.3 Imidazolidines	131
	3.4 Tetrahydroquinolines	131
	3.5 Tetrahydroquinazolines	133
	3.6 Dihydroquinazolines	134
	3.7 Pyrazolo hexahydroisoquinolines	135
	3.8 Azepines	136
	3.9 Lactams	140
	3.10 Ureas	144
4. Cycloaddition reactions for the formation of carbocycles		145
	4.1 Cyclopentanes	145
	4.2 Cyclopentenes	156
	4.3 Cyclohexanes	158
	4.4 Cyclohexenes	160
	4.5 Cycloheptanes	161
	4.6 Bicyclodecadienes	162
5. Cycloaddition reactions for the formation of mixed-heterocycles		163
	5.1 Oxazoline-related compounds	163

Advances in Catalysis, Volume 69
ISSN 0360-0564
https://doi.org/10.1016/bs.acat.2021.11.003

Copyright © 2021 Elsevier Inc.
All rights reserved.

103

5.2	Oxazepines	165
5.3	Epoxybenzoazepines	166
5.4	Oxazocines	167
5.5	Tetrahydrobenzoepoxyazocines	167
5.6	Oxazonines	168
5.7	Oxazecines	169
6. Conclusions		171
Acknowledgments		171
References		172
About the authors		180

Abstract

Asymmetric metal-catalyzed cycloaddition reactions via interceptive allylic substitution have become over the last decades a very appealing and powerful alternative methodology to classical cycloaddition reactions for the synthesis of chiral carbon- and heterocyclic compounds. Herein, the literature data reported since 2017 are compiled according the nature of the cyclic backbone formed.

1. Introduction

Metal-catalyzed cycloaddition reactions represents a very appealing alternative to classical cycloaddition reactions, such as the Diels-Alder reaction. Classical cycloaddition reactions are mainly governed by orbital symmetry considerations, which limits the type of cyclic synthons that can be accessed. To overcome this issue, the asymmetric metal-catalyzed cycloaddition reactions via interceptive allylic substitution has become an extremely powerful alternative methodology for the construction of a wide range of chiral carbon- and heterocyclic backbones not easily accessible using classical approaches (1). In this type of cycloaddition reactions, metal–zwitterionic species react with a dipolarophile (Scheme 1). Mechanistically, most of

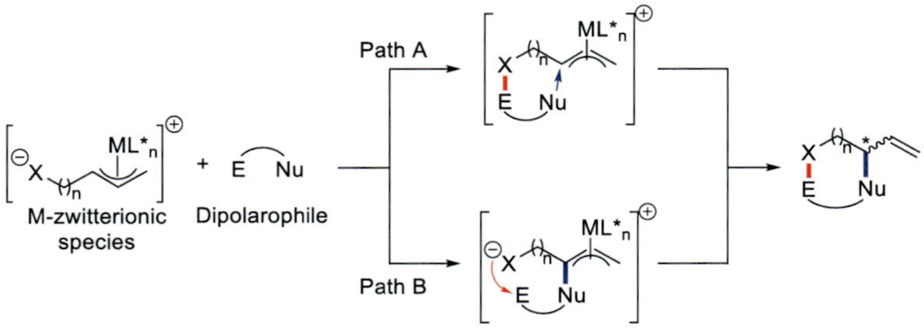

Scheme 1 Asymmetric metal-catalyzed cycloaddition reactions via interceptive allylic substitution.

the examples reported proceeds via a first nucleophilic attack of the metal-zwitterionic species to the dipolarophile followed by the nucleophilic attack of the dipolarophile to the metal-allyl fragment (path A; Scheme 1). In the last years, however, several successful reports have presented examples in which the dipolarophile firstly reacts as a nucleophile toward the metal-zwitterionic species (path B; Scheme 1).

Due to the prominent role of Pd-catalysts in allylic substitution reactions *(2)*, most of the reported examples make use of chiral Pd-catalyst, albeit in the last years the use of Ir-catalysts has also shown its potential utility in this transformation *(1)*. A wide assortment of allylic substrates able to generate metal-zwitterionic species can be employed, being vinylepoxides, vinylaziridines, vinylcyclopropanes and silylated allylic substrates the most commonly used. The range of dipolarophiles has also grown over time. Nowadays, the dipolarophile scope therefore includes activated alkenes as well as a wide range of unsaturated compounds (e.g., ketones, imines, aldehydes, etc.). In addition, recent works have shown that dipolarophiles can be also generated by organo- and photocatalysis *(3)*.

Herein, we compile the literature on this topic published during the last 5 years. Due to the diverse set of carbon- and heterocyclic compounds and having in mind the high potential of this transformation for the synthesis of highly complex organic molecules, we have grouped the literature data according the type of cyclic compounds formed.

2. Cycloaddition reactions for the formation of *O*-heterocycles

2.1 Tetrahydrofurans

One of the simplest ways to construct a furan ring via metal-catalyzed [3 + 2] cycloaddition is the reaction of either vinyl epoxides or vinylethylene carbonates with electron-deficient olefins *(4)*. Thus, for instance, Ding, Peng, Hou and coworkers disclosed the Pd-catalyzed [3 + 2] cycloaddition of a range of vinyl epoxides with several linear α,β-unsaturated enones (Scheme 2A) *(5)*. The ligand choice has a key impact in the success of the reaction. Thus, while catalytic systems modified with phosphine-oxazoline *i*-Pr-PHOX ligand, Feringa's phosphoroamidite and DACH-phenyl ligand proved to be unreactive, the use of Pd/(*R*)-BINAP catalyst gave the cycloaddition products in high yields, diastereo- and enantioselectivities. Nevertheless, long reaction times (typically 5 days) are required to full conversions due to less activated nature of the enones. Interestingly, while the structure of the epoxide didn't show any important effect on the stereochemical outcome of the reaction,

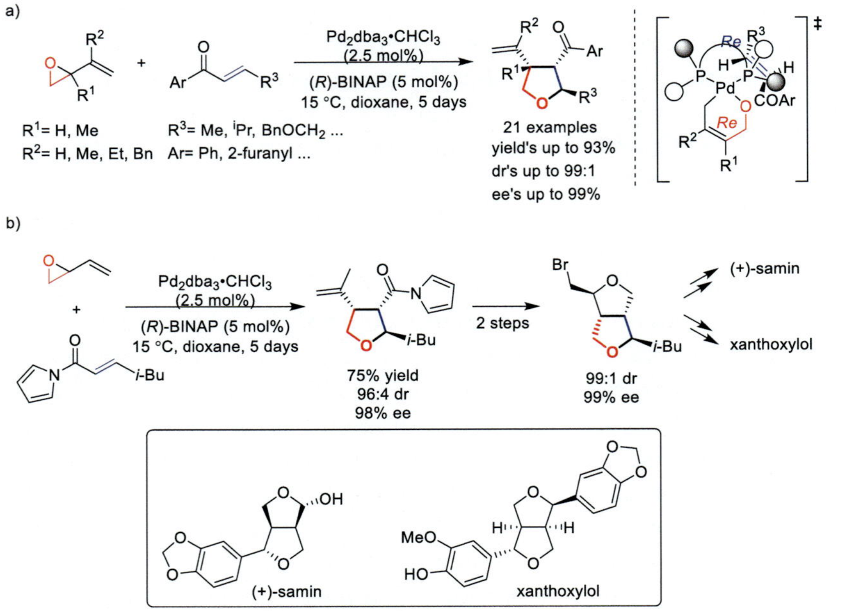

Scheme 2 (A) Pd/(R)-BINAP catalyzed [3+2] cycloaddition of vinyl epoxides with linear α,β-disubstituted enones. Representation of the key most stable *Re—Re* transition state that explains the observed stereocontrol. (B) Formal synthesis of (+)-samin and xanthoxylol.

the structure of the enone did. Thus, the use of ketones with no β-substituent led to lower diastereoselectivity. Similar decrease of selectivity was observed when using enones with *Z*-geometry. DFT calculations of the different transitions states (TSs) indicated that reaction between the *Re*-face of the Pd-zwitterionic adduct and the *Re*-face of the enone is more favorable than the rest of TSs (Scheme 2A), which fully explains the high selectivity achieved. The authors also demonstrated the usefulness of the methodology with the formal synthesis of natural furofuran lignans (+)-samin *(5)* and xanthoxylol (Scheme 2B) *(6,7)*.

The same year, You and coworkers reported the Pd-catalyzed dearomative [3+2] cycloaddition of several vinyl epoxides with a range of nitrobenzofurans (Scheme 3A) *(8)*. Again, the ligand choice proved to be crucial since catalysts containing DACH-phenyl ligand and Feringa's phosphoroamidite proved to be inactive. Nevertheless, the use of PHOX-type ligands gave the desired dearomatized cyclization. Among them, Pd-catalyst containing the electron-poor PHOX ligand (*S*)-**L1** gave the highest

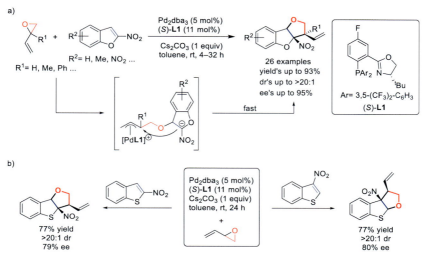

Scheme 3 Synthesis of (A) tetrahydrofurobenzofurans and (B) tetrahydrofurobenzothiophenes via Pd/(S)-**L1** catalyzed dearomative [3+2] cycloaddition of vinyl epoxides with nitrobenzofurans and nitrobenzothiophenes.

diastereo- and enantioselectivities (up to >20:1 dr and up to 95% ee; Scheme 3A). Interestingly, this methodology was also extended to substituted vinyl epoxides leading to tetrahydrofurobenzofurans with a vicinal chiral quaternary carbon stereocenter. Mechanistic investigations suggest that the zwitterionic Pd-allyl complex attacks the activated C-3 of the nitrobenzofuran, which is then followed by a fast cyclization (Scheme 3A). This methodology represents an interesting alternative for the synthesis of the tetrahydrofurobenzofuran core, which is present in many natural products and pharmaceuticals (e.g., (−)-panacene, (+)-psorofebrin, (+)-gynunone, etc.).

The same group extended this methodology to the dearomative cyclization of nitrobenzothiophenes (Scheme 3B) *(8)*. Under the same reaction conditions, the corresponding tetrahydrofurobenzothiophenes were attained in good yields and excellent diastereoselectivities. Albeit the enantioselectivities were somewhat lower to those achieved with related nitrobenzofurans (ee's up to 80%).

In 2018, You and coworkers, applied the same methodology used in previous nitrobenzofurans *(8)* to the [3+2] cycloaddition reaction of vinyl epoxides to nitroindoles *(9)*. Interestingly, the adequate choice of reaction conditions and ligand (S)-**L1** gave access to both diastereomers in high selectivities (up to >20:1 dr and up to 98% ee; Scheme 4A). Screening of several

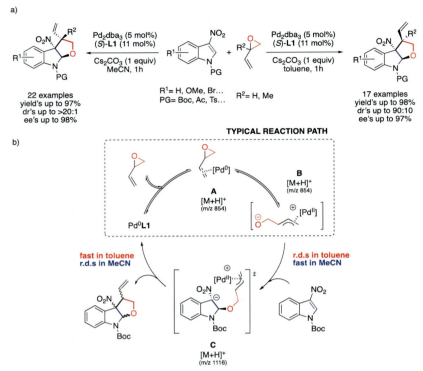

Scheme 4 (A) Synthesis of tetrahydrofuroindoles via Pd/(S)-**L1** the [3+2] cycloaddition reaction of vinyl epoxides to nitroindole**s**. (B) Proposed mechanism for the cycloaddition. In red the path using toluene as a solvent. In blue the path using MeCN as solvent.

protecting groups in the nitrogen atom showed that *t*-butyl carbonyl (Boc) group provided the best results, since other more electrodonating moieties like benzyl groups (Bn) didn't show any conversion under the same reaction conditions. Moreover, a diverse range of substituents in the aromatic substituent in the nitroindole substrate were successfully tested resulting in a broad substrate scope. The authors also proved the potential application of this reaction carrying out several transformations on a selected product without erosion of the stereometric purity *(9)*. Mechanistic studies using SAESI-MS detected the presence of key intermediates of the reaction **A**, **B** and **C** (Scheme 4B) in both toluene and acetonitrile. These studies together with kinetic studies (carried out by ^{1}H NMR and HPLC) aided the authors to propose a possible catalytic cycle in which the rate-determining steps varies depending on the solvent. In toluene (red in Scheme 4B), the rate-determining step is the dearomative Michael addition to form **C**, which is then followed by a fast cyclization. The slow Michael addition in toluene is due to the formation of a thigh ion pair in nonpolar solvents, which decreases the nucleophilicity of the Pd–zwitterionic intermediate **B**. On the other hand, in acetonitrile (blue in Scheme 4B), intermediate **C** is easily formed but the following ring-closing step becomes the rate-determining step due to the stabilization of the Pd–zwitterionic intermediate **C** by the polar solvent. This is the key step in the solvent-dependent diastereomer formation, since this stabilization provides time for π–σ–π inversion of the allylic palladium complex that is required to attain the other diastereoisomeric compound.

In the same year, Zhang and coworkers reported the decarboxylative [3+2] cycloaddition of vinylethylene carbonates to monoactivated β-nitroolefins *(10)*. This methodology takes advantage of the presence of chiral squaramides in order to obtain high enantioselectivities. In this case, the mechanism proceeds similarly as in previous examples, where an oxo-Michael addition of a zwitterionic palladium complex to the nitroolefins takes place. The presence of the chiral squaramide activates the nitroolefins by means of hydrogen bonding and, at the same time, has a synergistic effect with the chiral Pd-catalyst on the stereochemical outcome of the reaction (Scheme 5A). Optimization experiments proved that chiral phosphoroamidite

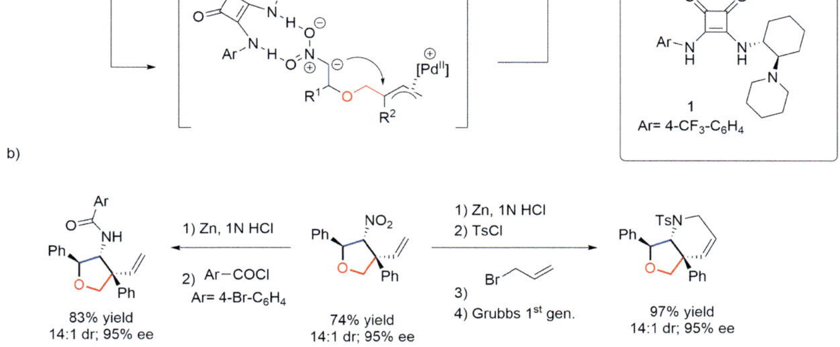

Scheme 5 (A) Dual organocatalysis and Pd-catalyzed [3+2] cycloaddition of vinylethylene carbonates with linear β-nitroolefins. Representation of the key intermediate involving hydrogen bonding interactions with the squaramide organocatalyst. (B) Derivatization of benchmark nitrotetrahydrofurans.

(*S*)-**L2** in combination with chiral squaramide **1** provided the best results (up to >20:1 dr and up to 98% ee; Scheme 5A). The synthetic versatility of this methodology was demonstrated by the gram scale synthesis and derivatization of benchmark nitrotetrahydrofurans into amides and bicyclic heterocycles (Scheme 5B).

Interestingly, also in 2018, Hou and coworkers reported a very similar [3+2] cycloaddition of vinyl epoxides with monoactivated β-nitroolefins *(11)*. In this case, no external activation was utilized, yielding nitrotetrahydrofurans in high diastero- and enantioselectivities when using (*R*)-BINAP as a ligand (up to 20:1 dr and up to 99% ee; Scheme 6). Mechanistic studies indicate that there is a possible isomerization of the *Z*-isomer to the *E*-isomer through a reversible oxo-Michael addition; hence, the geometry of the olefins is crucial to achieve high enantioselectivities. DFT calculations also showed that the addition of the *Re*-face of the vinyl epoxide to the *Re*-face of the zwitterionic Pd complex is the most favored pathway. The calculations are in agreement with the absolute configurations of the final products.

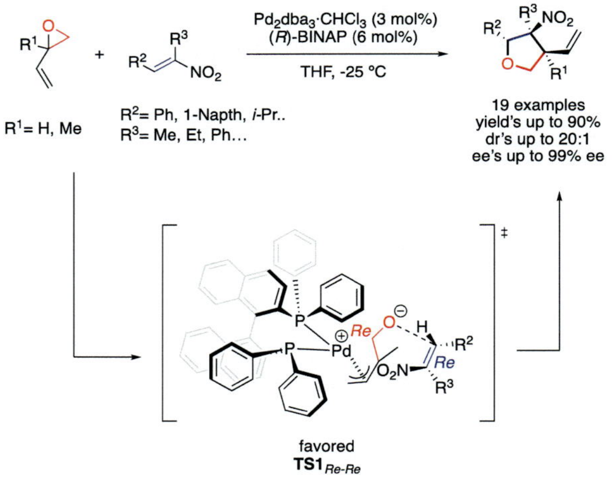

Scheme 6 Pd/(*R*)-BINAP catalyzed [3+2] cycloaddition of vinyl epoxide with monoactivated β-nitroolefins.

Soon later, Zhang and coworkers developed an efficient method for the construction of furanobenzodihydropyran skeletons using the decarboxylative [3+2] cycloaddition of vinylethylene carbonates to activated chromones (Scheme 7) *(12)*. After optimization, phosphoroamidite ligand (*S,R,R*)-**L3** proved to yield good diastereoselectivities and high enantioselectivities (up to 5:1 dr and up to 97% ee; Scheme 7). Gram scale transformation and

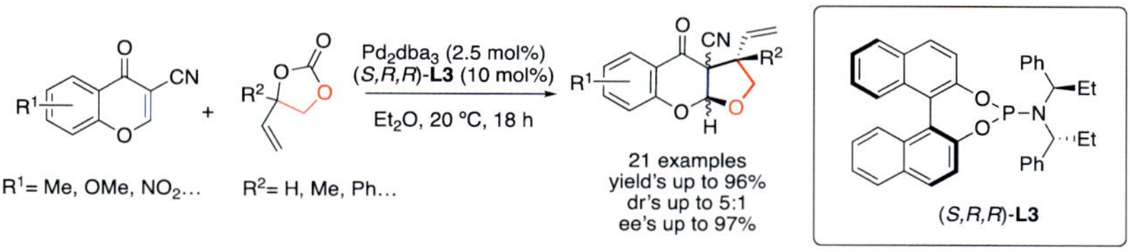

Scheme 7 Pd/(S,R,R)-**L3** catalyzed [3+2] cycloaddition of vinylethylene carbonate with 2-cyanochromanones.

derivatization of the keto group into the corresponding chiral alcohol was performed in order to demonstrate the versatility of the procedure *(12)*.

A new approach for the synthesis of alkylidene derived tetrahydrofurans was disclosed by Hou and coworkers that applied the [3+2] cycloaddition of methyl-substituted vinyl epoxide with α-allenamides (Scheme 8A) *(13)*. Best results were obtained with a NHC-based ligand (S,R)-**L4** reaching high - diastereo- and enantioselectivities (up to 20:1 dr and up to 97% ee; Scheme 8A). Nevertheless, conversions are highly dependent on the nature of the vinyl epoxide employed. Thus, whereas high yields were attained using 2-methyl-2-vinyloxirane, the use of other vinyl epoxides led to low conversions (Scheme 8B). It should be also noted the no conversions were observed using 4-unsubstituted, 4-monosusbtituted, 4,4,-diphenyl, 1,4,4-trisubtituted, N,N-diethyl allenic amides and α-allenic esters as substrates *(13)*.

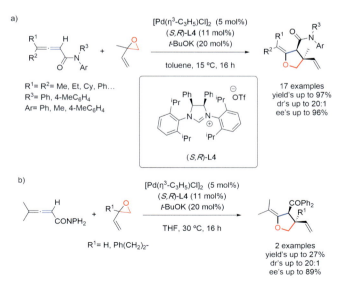

Scheme 8 (A) Pd/(S,R)-**L4** catalyzed [3+2] cycloaddition of 2-methyl-2-vinyloxirane with α-allenamides. (B) Reaction of oher vinyl epoxides with benchmark α-allenamide.

In 2019, Guo and coworkers found a very convenient way to efficiently synthetize isonucleoside analogs via [3 + 2] cycloaddition of 2-vinyloxirane to α–N-heterocyclic acrylates *(14)*. PHOX ligand (S,S_P)-**L5** yielded the best results (up to >20:1 dr and up to 98% ee; Scheme 9A) for a broad range of α–N-heterocyclic acrylates. However, with α-pyrimidine-substituted acrylates a change of conditions and ligand to (S)-**L6** were necessary to obtain the best results (up to 7:1 dr and up to 95% ee; Scheme 9B) *(14)*.

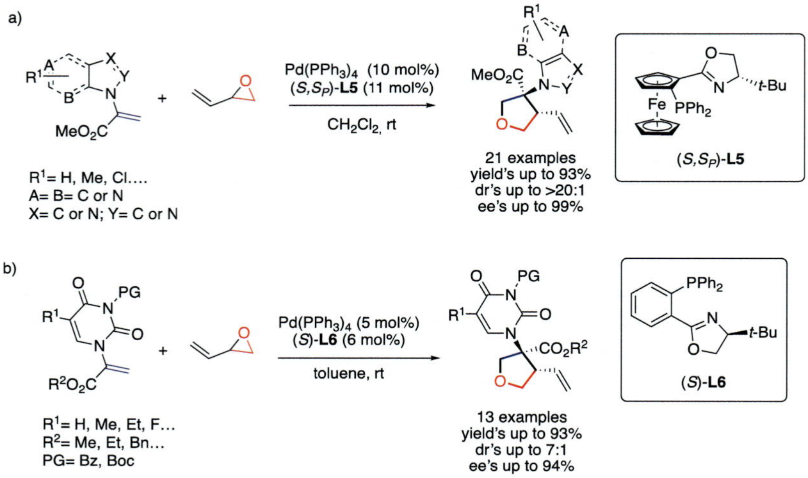

Scheme 9 Pd-catalyzed [3 + 2] cycloaddition of 2-vinyloxirane with (A) α-N-heterocyclic acrylates and (B) α-pyrimidine-substituted acrylates.

The next year, Hu and coworkers developed the synthesis of spiro-oxindoles through a decarboxylative [3 + 2] cycloaddition of vinylethylene carbonates to methylene indolinenones *(15)*. After reaction optimization, high diastereo- and enantioselectivities were achieved (up to >20:1 dr and up to 99% ee; Scheme 10) using diphosphine (S,S,S,S)-**L7**. The robustness

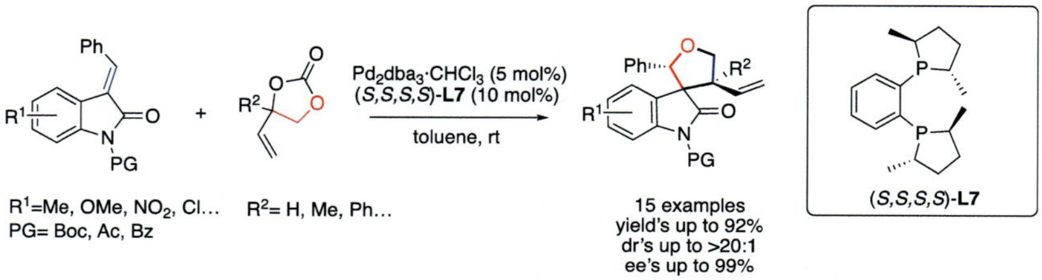

Scheme 10 Pd/(S,S,S,S)-**L7** catalyzed [3 + 2] cycloaddition of vinyl epoxides to a range of indolinenones.

of the synthetic protocol was further demonstrated by the gram–scale preparation of the cycloadducts and several post-functionalizations *(15)*.

Also in 2020, Guo and coworkers presented in a similar manner an efficient synthesis of spirocyclic 1,3-indandiones *(16)*. Thus, the decarboxylative cycloaddition of vinylethylene carbonate derivatives with several benzylidene 1,3-indandiones employing phosphoroamidite ligand (R,S,S)-**L8** led high diastereo- and enantioselectivities (up to 4:1 dr and up to 99% ee; Scheme 11).

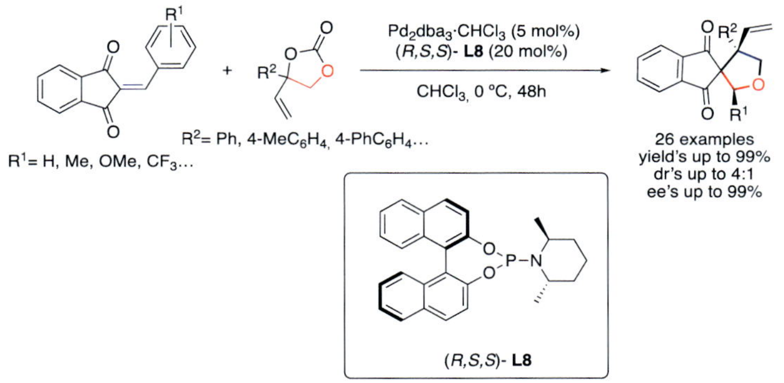

Scheme 11 Pd/(R,S,S)-**L8** catalyzed [3 + 2] cycloaddition of vinylethylene carbonates with benzylidene 1,3-indandiones.

Later that same year, Zhang and coworkers developed a Pd-catalyzed [3 + 2] cycloaddition of vinylethylene carbonates on double activated 2-nitroacrylates *(17)*. (S)–SegPhos proved to be the best ligand reaching high diastereo- and enantioselectivities (up to 20:1 dr and up to >99% ee; Scheme 12A). As in previous examples *(5)*, these tetrahydrofuran scaffolds

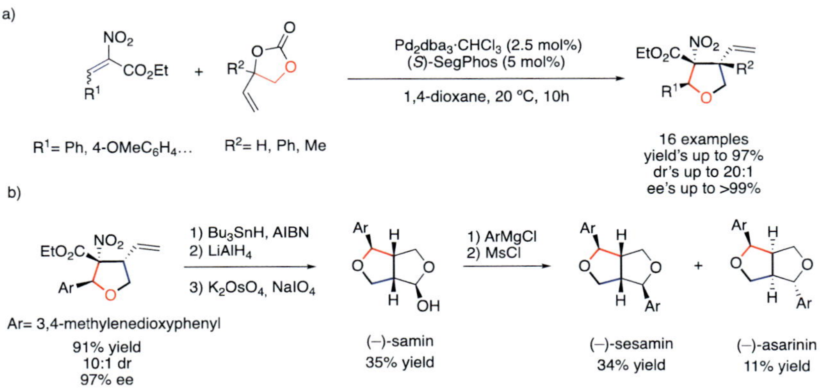

Scheme 12 (A) Pd/(S)-SegPhos catalyzed [3 + 2] cycloaddition of vinylethylene carbonates with 2-nitroacrylates. (B) Formal synthesis of lignans (−)-samin, (−)-sesamin and (−)-asarinin.

were used in the formal synthesis of highly valuable lignans: (−)-samin, (−)-sesamin and (−)-asarinin (Scheme 12B).

Another relevant methodology for the synthesis of chiral tetrahydrofuran rings is the reaction of activated vinyl cyclopropanes with an O–containing heterodipolarophile. In this context, Yang and coworkers used the vinyl cyclopropane approach but, in this case, in an intermolecular fashion. Thus, bispirooxindole **2** was prepared via [3 + 2] cycloaddition of spirovinyl-cyclopropyl oxindole with 1-benzylindoline-2,3-dione using Pd/(*S,S,S*)-**L9** catalytic system (Scheme 13) *(18)*.

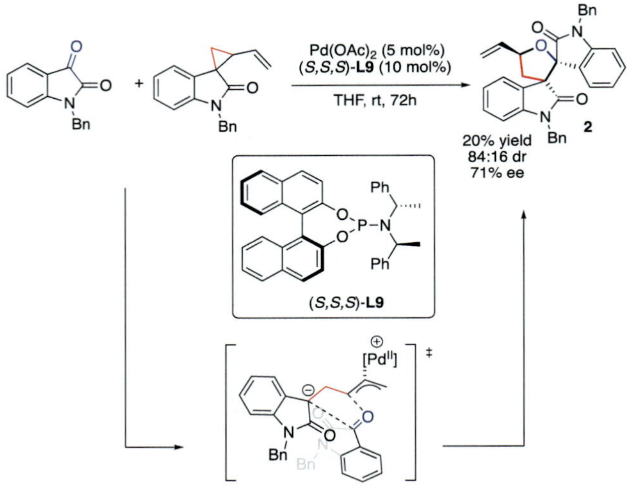

Scheme 13 Synthesis of bispirooxindole **2** via Pd/(*S,S,S*)-**L9** catalyzed [3 + 2] cycloaddition.

The vinyl cyclopropane approach has been also used in combination with other type of substrates. In 2019, Xiao and coworkers reported the *in situ* photoinduced Wolff rearrangement and subsequent Pd-catalyzed [3 +2] cycloaddition of activated cyclopropanes to α–diazoketones *(19)*. The key step in this tandem reaction is the formation of a ketene intermediate via photoinduced Wolff rearrangement and then, upon addition of the Pd–catalyst precursor, ligand and substrate; the Pd–catalyzed [3 + 2] cycloaddition takes place (Scheme 14). However, only one asymmetric example was reported. The use of P,S ligand (*S,R*)-**L10** therefore provided the cyclic adduct in 91% yield and 83% ee. Interestingly, soon latter Kerringan and coworkers reported a similar approach for the synthesis of tetrahydrofurans with an exocyclic double bond starting from the preformed ketene *(20)*.

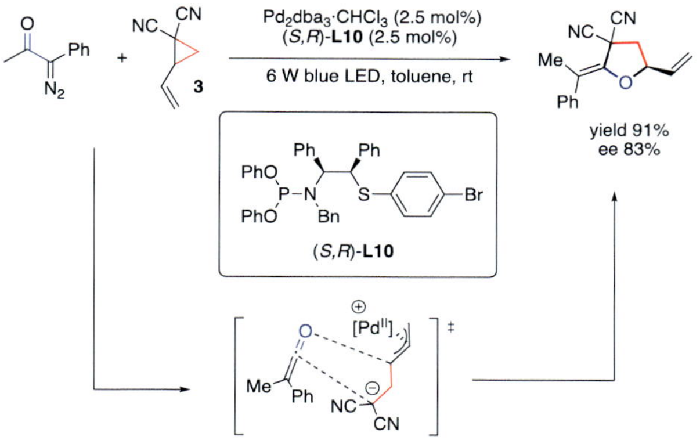

Scheme 14 Tandem Wolff rearrangement/ Pd/(*S*,*R*)-**L10** catalyzed [3 + 2] cycloaddition of vinyl cyclopropane **3** with 1-diazo-1-phenylpropan-2-one.

Nevertheless, no asymmetric version was reported since only enantiopure vinyl cyclopropanes were tested.

Another strategy that can be employed in the construction of chiral tetrahydrofurans is the use of the Pd-trimethylenemethane (TMM) methodology developed in Trost's group *(1e)* in combination with O-dipolarophiles. Thus, in 2020, Trost and coworkers used the Pd–TMM approach to synthetize trifluoromethylated tetrahydrofuran scaffolds. In this case the *in situ* generated Pd–TMM zwitterion reacts with a carbonyl-type compound in a [3 + 2] fashion to yield the corresponding polyfluorinated-tetrahydrofuran *(21)*. Reaction of electron deficient aldehydes or trifluoromethyl–substituted 2-acetonaphthone afforded trifluoromethylated tetrahydrofurans in excellent selectivities using diamidophosphite (*S,S,R,R,S,S*)-**L11** (up to 9:1 dr and up to 97% ee; Scheme 15A). It should be mentioned that simpler ketones like acetophenone and acetone did not afford any product. In addition, a series of *N*-methyl isatins were tested under cycloaddition conditions yielding tetrahydrofurans with a spiro quaternary center in high diastereo- and enantioselectivity (up to 8:1 dr and up to 99% ee; Scheme 15B). Additionally in the isatin–derived substrates, the CF_3 group could be substituted by a CHF_2 without erosion of the selectivity (3:1 dr and 98% ee; Scheme 15B).

2.2 Dihydrofurans

Tetrasubstituted dihydrofurans with a trifluoromethylated group are found in many bioactive molecules *(22)*. To achieve such compound type, Trost and coworkers disclosed a methodology based in the [3 + 2] cycloaddition of *in situ* formed Pd–TMM zwitterion with fluorinated ketones *(23)*. They found that the use of Boc substituted nitrile-methyallyl donor **4** is key to

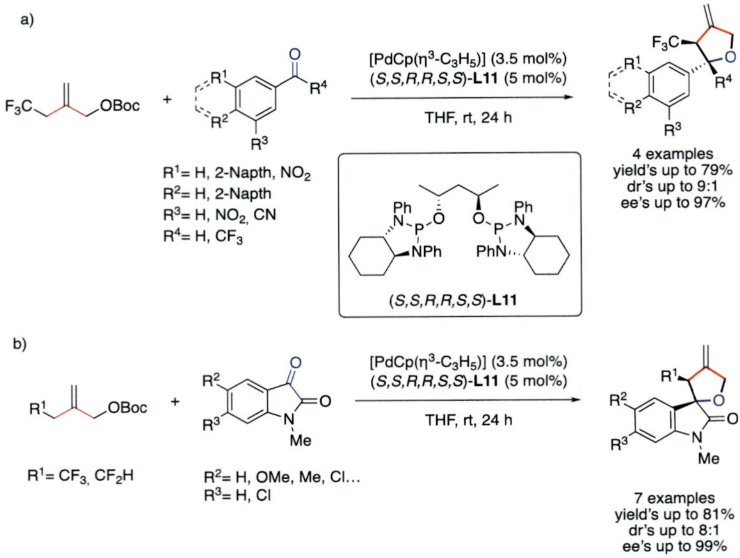

Scheme 15 Pd/(S,S,R,R,S,S)-**L11** catalyzed [3 + 2] cycloaddition of polyfluorynated-methyallyl donnors to a) aldehydes or ketones and b) N-methyl isatins.

achieve total control of the regioselectivity (>20:1 *endo*). High selectivities were therefore obtained regardless of the nature of the substituents of the ketone using phosphoroamidite (R,R,R,S_P)-**L12** (up to 99% ee; Scheme 16). This methodology was also used to synthetize a variety of interesting interme-diates by derivatization of the nitrile group.

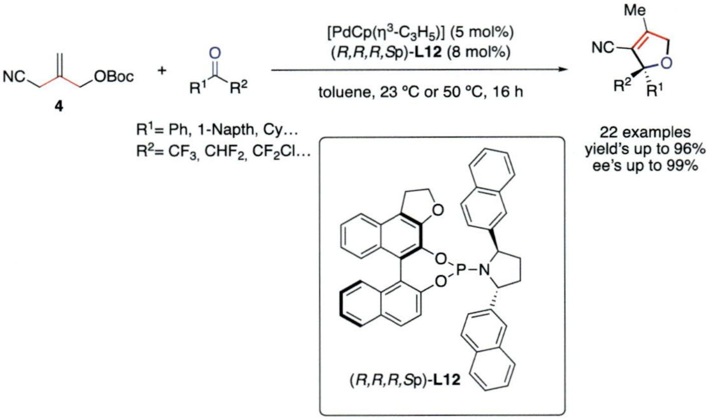

Scheme 16 Pd/(R,R,R,Sp)-**L12** catalyzed [3 + 2] cycloaddition of **4** with fluorinated ketones.

2.3 Tetrahydropyrans

Similar to the construction of tetrahydrofurans, the synthesis of chiral tetrahydropyrans can be achieved by asymmetric cycloaddition reactions involving a Pd-zwitterionic species. For example, Guo and coworkers recently reported the synthesis of tetrahydropyran-fused spirocyclic compounds via asymmetric [4 + 2] cycloaddition of 2-methylidenetrimethylene carbonate with activated exocyclic alkenes (24). Thus, a range of spiropyrazolones were synthesized in excellent yields and enantioselectivities using Segphos derivative (R)-**L13** (up to 99% ee, Scheme 17A). Furthermore, this approach was efficiently extended to the use of other types of activated alkenes such as those derived from indandiones (up to 99% ee; Scheme 17B) and barbiturates (up to 99% ee; Scheme 17C).

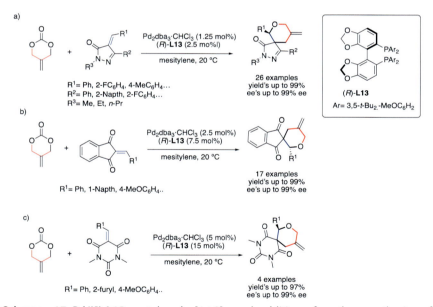

Scheme 17 Pd/(R)-**L13** catalyzed [4 + 2] cycloaddition for the synthesis of (A) spiropyrazolones, (B) spiroindandiones and (C) spirobarbiturates.

In the same year, Hou and coworkers reported the synthesis of simple tetrahydropyrans taking advantage of the 1,4 dipole Pd-catalyzed [4 + 2] cycloaddition of 4-vinyl-1,3-dioxan-2-ones with α,β-disubstituted nitroalkenes (25). For this transformation, PHOX derived ligand (S,S)-**L14** provided the best diastereo- and enantioselectivities (up to 20:1 dr and up to 98% ee; Scheme 18). The authors also indicated that Z/E isomerization

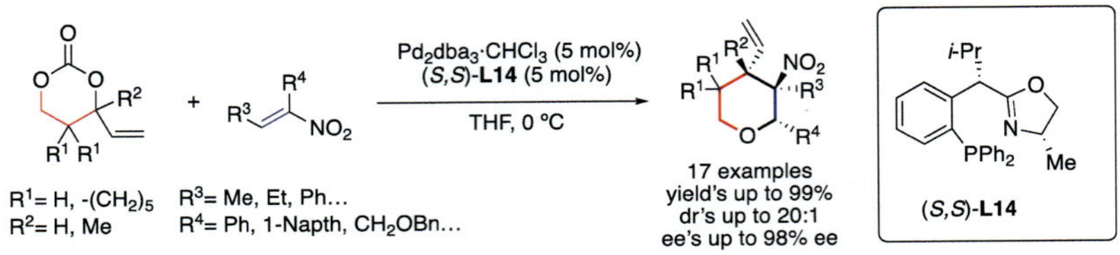

Scheme 18 Pd/(S,S)-**L14** catalyzed [4+2] cycloaddition for the synthesis of highly substituted tetrahydropyrans.

of the alkene takes place in situ through a reversible oxo–Michael addition of the Pd-alkoxide. As a results, the geometry of the nitro olefin has a significant effect on the outcome of the reaction. Thus, the use of Z-nitro olefins led to lower yields and selectivities than the E-nitroalkenes. Additional post functionalization of the cycloadducts was performed to demonstrate the usefulness of the protocol *(25)*.

2.4 Dioxanes

Chiral 1,3-dioxane scaffolds can be easily transformed in 1,3-diols which are present in many pharmaceutical and natural products *(26)*. In 2019, Zhang and coworkers, developed an efficient way to construct 1,3-dioxanes through the formal Pd-catalyzed [4 + 2] cycloaddition of vinyl oxetanes with formaldehyde using phosphoroamidite ligand (S,R,R)-**L3** (Scheme 19) *(27)*. Under the same reaction conditions, the use of cyclic carbonates, such as 4-phenyl-4-vinyl-1,3-dioxan-2-one, proved to be good alternatives for the vinyl oxetanes, providing similar high yields and ee's. The versatility of the synthetic protocol was demonstrated by the elaboration of enantio-enriched 1,2,4-triol from the 1,3-dioxane derivative in three easy steps without erosion of the enantioselectivity *(27)*.

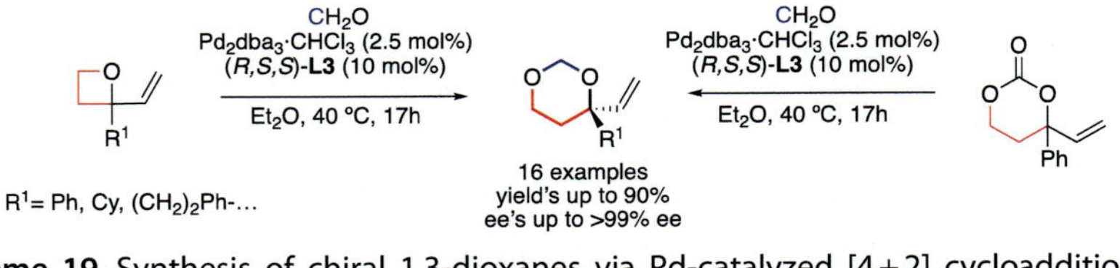

Scheme 19 Synthesis of chiral 1,3-dioxanes via Pd-catalyzed [4+2] cycloaddition of vinyl oxetanes and 4-phenyl-4-vinyl-1,3-dioxan-2-one with formaldehyde.

2.5 Lactones

Medium sized ring lactones represent a very interesting synthon for organic synthesis *(28)*. One of the most convenient pathways to synthetize lactones in one step is through metal-catalyzed dipolar cycloadditions. For example, in 2018, Glorius and coworkers reported a Pd–catalyzed [5 + 2] annulation of vinylethylene carbonates with α,β-unsaturated aldehydes assisted by N–heterocyclic carbenes (NHC) as organocatalyst *(29)*. A range of 7-membered lactones were therefore attained in excellent diastereo- and enantioselectivities using (R)-BINAP derivate (R)-**L15** and chiral imidazolium salt precatalyst **5** (up to 50:1 dr and up to >99% ee; Scheme 20). Mechanistic investigations revealed that the presence of hydrogen bonding between the homoenolate-NHC adduct and the Pd-zwitterionic complex is crucial in the success of this [5 + 2] cycloaddition reaction.

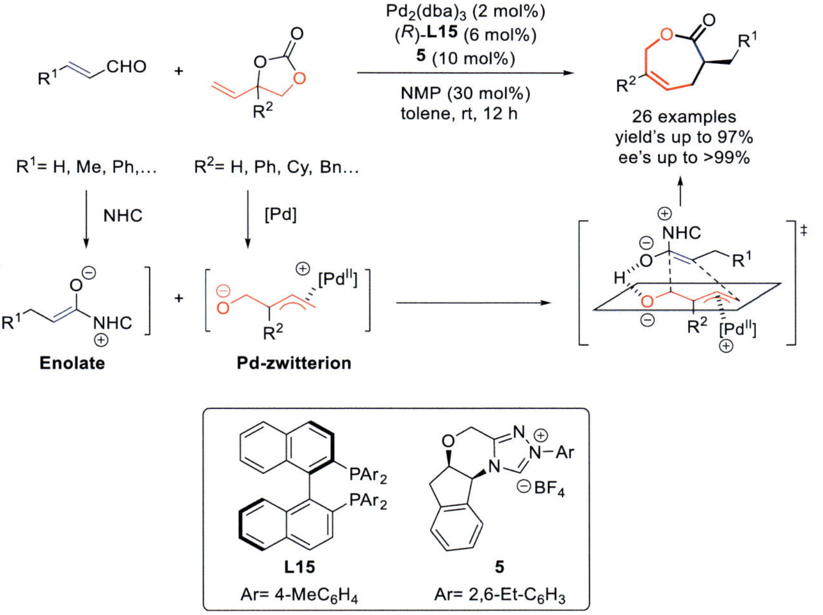

Scheme 20 Synthesis of 7-membered ring lactones via [5 + 2] cycloaddition of vinylethylene carbonates to α,β-unsaturated aldehydes through cooperative carbene organocatalysis and palladium catalysis.

More recently, the same group made use of the dual NHC-based organocatalysis and metal-catalyzed cycloaddition for the synthesis of five-membered ring lactones *(30)*. A range of α,β-disubstituted γ-butyrolactones were efficiently prepared using Ir/(S)-**L16** catalytic system in combination with N-heterocyclic carbene precursor **6** (up to >20:1 dr and up to >99% ee; Scheme 21A). In this case, the Ir-zwitterion prepared from the decarboxylation of vinylethylene carbonate underwent cycloaddition with the

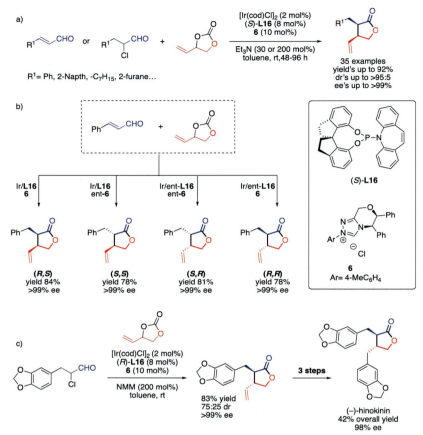

Scheme 21 (A) Synthesis of γ-butyrolactones via cooperative NHC organocatalysis and Ir-catalyzed cycloaddition of vinylethylene carbonate with α,β-unsaturated aldehydes or α-chloro aldehydes. (B) Diastereodivergent synthesis of all possible isomers of 3-phenyl-4-vinyldihydrofuran-2(3H)-one. (C) Synthesis of (−)-hinokinin.

desired azolium–enolate prepared from the corresponding α,β-unsaturated aldehyde and **6** in the same manner as depicted in Scheme 20. It is to note that the use of polyaromatic and heteroaryl-containing aldehydes required longer reaction times. To overcome this drawback the authors replaced the α,β-unsaturated aldehyde by a more reactive α-chloroaldehyde derivative (Scheme 21A). The versatility of this protocol was proved by the synthesis of all four isomers of a benchmark γ-butyrolactone by means of a diastereodivergent [3 + 2] cyclization (Scheme 21B). Moreover, the formal synthesis of lignan (−)-hinokinin was performed further demonstrating the effectiveness of the catalysis protocol (Scheme 21C) *(30)*.

Later, Xiao and coworkers took a different approach to build medium sized lactones. In this case, they reported the synthesis of 7-membered ring lactones from α-diazoketones and vinyl ethylene carbonates via in situ photoinduced Wolff rearrangement and subsequent Pd-catalyzed [5 + 2] cycloaddition *(31)*. As previously mentioned, the Pd-zwitterion reacts with the photogenerated ketene in a [5 + 2] fashion. After optimization, Pd/(*R*)-**L17** catalytic system proved to be highly efficient for a range of aryl-substituted vinylethylene carbonates and aryl-substituted α-diazoketones (up to >98%; Scheme 22). Kinetic studies show a positive nonlinear effect between the ee of the ligand and that of the cycloadduct *(31)*. Moreover, mechanism insight was gained with the elucidation of key intermediates through ESI-MS experiments and with the help of DFT calculations. The latter showed a steric match of the π-allyl fragment, the ketene and the chiral ligand that favors the *Si*-face ring closing (Scheme 22).

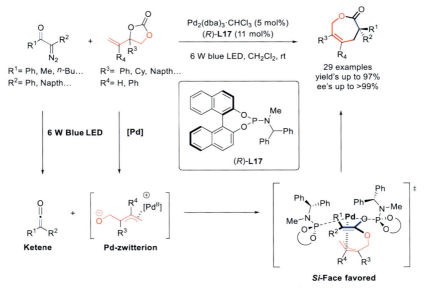

Scheme 22 Synthesis of 7-membered ring lactones via tandem Wolff rearrangement/ [3 +2] cycloaddition of vinyl cyclopropanes with α-diazoketones.

A year later, Xiao and coworkers used the same methodology to attain 7-membered lactone-fused polycyclic compounds via [5 + 2] asymmetric cycloaddition of α-diazoketones with 1,3-indandione-derived vinyl cyclopropanes *(32)*. A key in this transformation is that the cyclopropane ring opening by Pd led to the formation of the oxo-1,5-dipole rather than the

1,3-dipole due to the presence of keto group adjacent to the cyclopropane group. Then, reaction of these 1,5-dipoles with the desired photogenerated ketene yielded 7-membered lactone-fused polycyclic compounds in high diastereo- and enantioselectivities (Scheme 23) using P,S ligand (R)-**L18**.

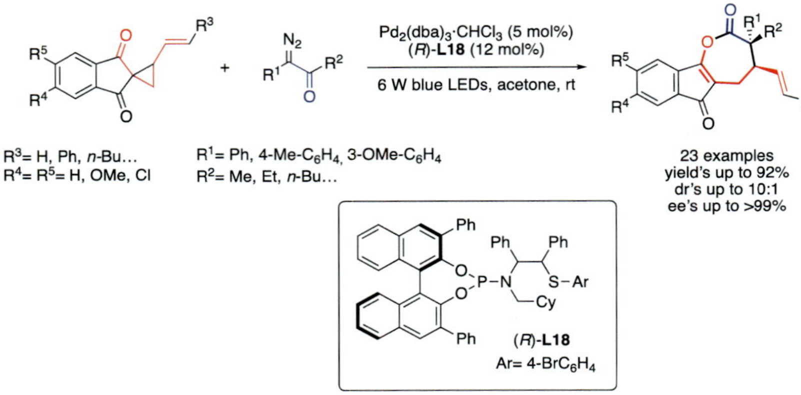

R³= H, Ph, n-Bu...
R⁴= R⁵= H, OMe, Cl

R¹= Ph, 4-Me-C₆H₄, 3-OMe-C₆H₄
R²= Me, Et, n-Bu...

23 examples
yield's up to 92%
dr's up to 10:1
ee's up to >99%

(R)-**L18**
Ar= 4-BrC₆H₄

Scheme 23 Synthesis of 7-membered lactone-fused polycyclic compounds via light driven [5 + 2] cycloaddition of α-diazoketones to 1,3-indandione-derived vinyl cyclopropanes uding Pd/(R)-**L18** catalytic system.

Another methodology for the synthesis of medium cyclic lactones was disclosed by Guo and coworkers *(33)*. The synthesis of 9-membered ring bicycle[5.2.2]tetrahydrooxonines were attained via the tandem Pd-catalyzed [3 + 2] cycloaddition/Cope rearrangement of vinylethylene carbonates with coumalates or pyrones (Scheme 24). The use of phosphoroamidite (R)-**L19** provided the highest enantioselectivities for the reactions with coumalates (up to 99% ee), whereas for reactions with pyrones, the highest ee's were attained using phosphoroamidite (R,S,S)-**L9** (up to 97% ee). It should be highlighted that for enantioselectivities to be high the use of (hetero) aryl-substituted vinyletylene carbonates was required. Gram scale synthesis and post derivatizations were performed to prove the potential utility of this new methodology *(33)*. Experimental and DFT calculations clearly indicated that the Pd-zwitterion underwent typical oxo–Michael addition to the methyl coumalate to yield the [3 + 2] cycloadduct, which then undergoes Cope rearrangement to yield the bicycle[5.2.2]tetrahydrooxonines rather than via a direct [5 + 4] cycloaddition. Calculations confirmed that the rate-determining step of the reaction is the [3 + 2] annulation *(33)*.

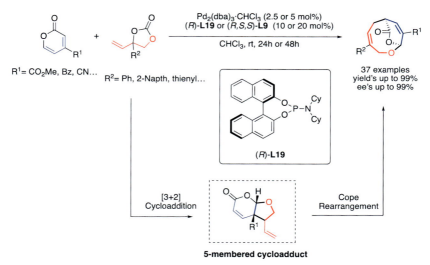

Scheme 24 Synthesis of 9-membered tetrahydrooxonine-fused lactones via tandem [3+2] cycloaddition/Cope rearrangement of vinylethylene carbonates with coumalates or pyrones.

2.6 Oxepines

Medium sized rings, especially 7-membered rings containing an heteroatom, are a very common recurrent motive in natural products and bioactive compounds *(34)*. In this context, Trost and coworkers recently disclosed the synthesis of benzo[*b*]oxepines via [4+3] cycloaddition of a range of *ortho*-quinomethides via the Pd-TMM zwitterion, formed from *tert*-butyl (3-((diphenylmethylene)-amino)prop-1-en-2-yl) carbonate. The use of Pd/(*R,S,S*)-**L9** proved to be optimal for this transformation providing high regio-, diastereo- and enantioselectivities (Scheme 25) *(35)*. It should be highlighted that no subproducts of the [6+3] cycloaddition were observed when the styryl-substituted *ortho*-quinone methides were used. Benzo[*b*]

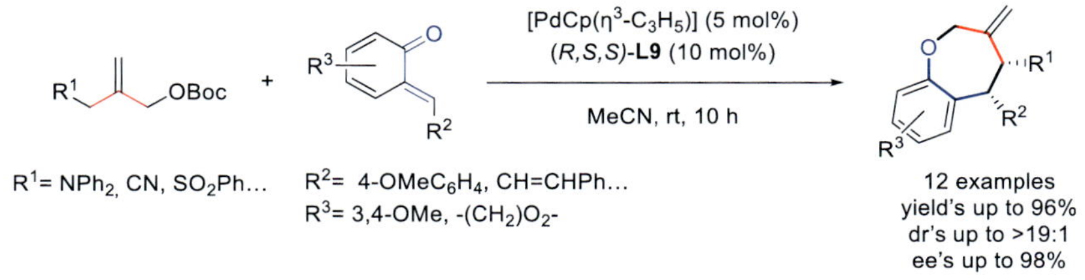

Scheme 25 Pd-TMM catalyzed [4+3] cycloaddition of *ortho*-quinone methides.

oxepines containing other functional groups adjacent to the exocyclic double bond rather than the imine group were also attained in similar high selectivities (up to >19:1 dr and up to 98% ee; Scheme 25).

2.7 Dioxonines

In 2020, Fan and coworkers, developed a new synthetic pathway to obtain 9-membered dioxonines via Pd-catalyzed [4+5] cycloaddition of *ortho*-quinone methides to substituted vinylethylene carbonates *(36)*. The use of Pd/(R)-BINAP catalytic system yielded de 9-membered ring dioxonines with high yields and enantioselectivities (up to 99% ee; Scheme 26A). One interesting feature of the formed 9-membered dioxonines is that they can be easily transformed to corresponding chiral phenol-containing homoallylic alcohols via Claisen rearrangement with a mild acid (e.g., silica gel) and sequential hydrolysis (Scheme 26B).

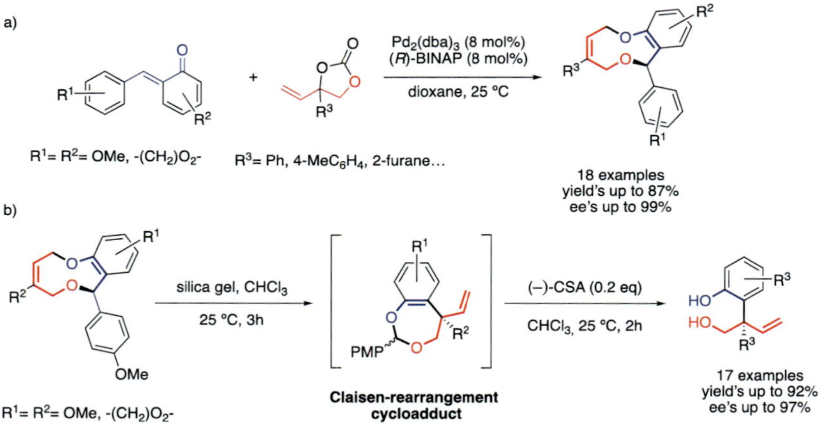

Scheme 26 Synthesis of (A) 9-membered dioxonines via Pd/(R)-BINAP catalyzed [4+5] cycloaddition of *ortho*-quinone methides to substituted vinylethylene carbonates and (B) phenol-functionalized homoallylic alcohols via tandem Claisen rearrangement/hydrolysis of 9-membered dioxonines.

3. Cycloaddition reactions for the formation of *N*-heterocycles

3.1 Pyrrolidines

Nitrogen heterocycles, especially five-membered rings, are recurring structural motifs in natural products and pharmaceuticals *(37)*. The Pd-catalyzed [3+2] cycloaddition of vinyl aziridines to electron deficient substrates has been the synthetic pathway of choice in recent years *(38)*. Thus, for example, Jørgensen and coworkers, developed a synergistic Pd/organocatalytic [3+2] cycloaddition of vinyl aziridines to α,β-unsaturated aldehydes *(39)*. In this procedure, the Pd-precursors facilitates the opening of the vinyl aziridine while the pyrrole-based organocatalyst is responsible of the enantio-induction. After optimization silyl protected compound **7** was the organocatalyst of choice to obtain high diastereo- and enantioselectivities the cycloaddition of vinyl aziridine **8** with a range of α,β-unsaturated aldehydes (up to >20:1 dr and up to >99% ee; Scheme 27A). They also found out that the use of the substituted vinyl aziridine substrate **9** slowed down the reaction significantly. In order to speed up the process, the use of racemic phosphoroamidite ligand **L20** was needed, albeit selectivity decreased (up to 11.5:7.8:1 dr and up to 40% ee; Scheme 27B). *In situ* post functionalization was performed to prove the synthetic viability of this procedure *(39)*.

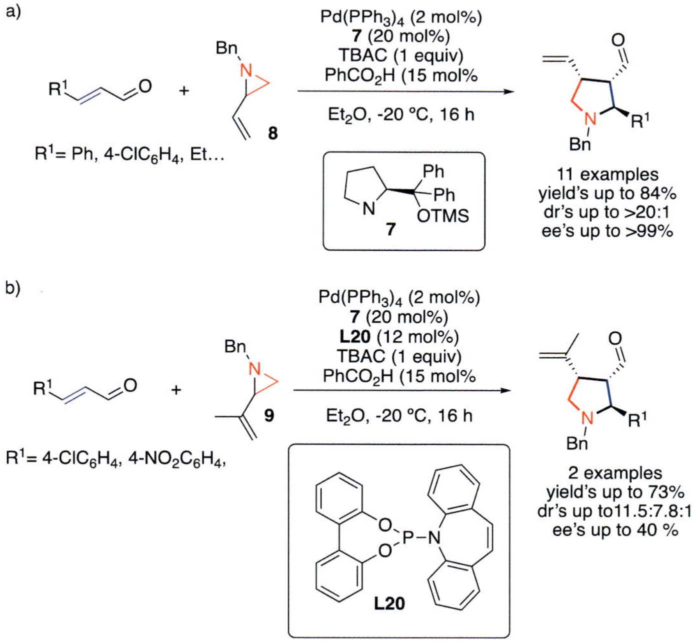

Scheme 27 Synthesis of chiral pyrrolidines via cooperative Pd/**7** [3+2] cycloaddition of a range of α,β-unsaturated aldehydes with (A) vinyl aziridine **8** and (B) vinyl aziridine **9**.

Later the same year, Glorius and coworkers reported a dual Pd/NHC [4+1] cycloaddition reaction of vinyl benzoxazinanones with sulfur ylides *(3a)*. NHC precatalyst **5** was selected as optimal ligand for this transformation reaching perfect diastereoselectivities and high enantioselectivities in almost all cases (up to >20:1 dr and up to 90% ee; Scheme 28).

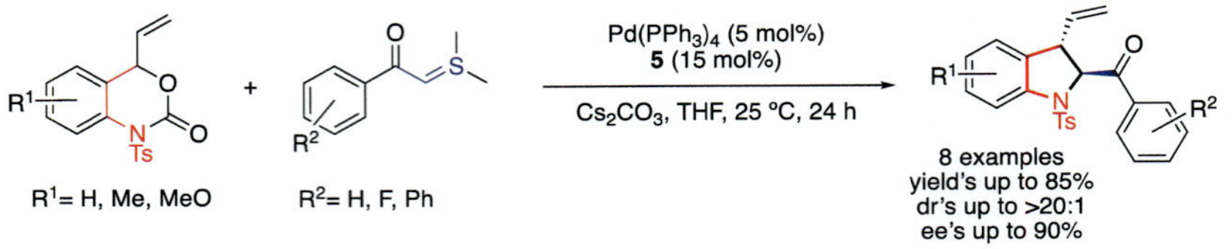

Scheme 28 Synthesis of acyl-pyrrolidinines via dual Pd/NHC [4 + 1] cycloaddition reaction of vinyl benzoxazinanones with sulfur ylides.

In 2018, Wang and coworkers reported a dearomative Pd-catalyzed [3 + 2] cycloaddition reaction between vinyl aziridine **10** and 3-nitroindoles to yield 3a-aminopyrroloindolines *(40)*. Ligand screening resulted in the finding of bisoxazoline (S,R,S,R)-**L21** as the ligand that provided the highest diastereo- and enantioselectivities in a range of 3-nitroindoles with tosyl protected vinyl aziridine **10** (up to >99:1 dr and up to 97% ee; Scheme 29). Reduction of the nitro moiety, hydroboration and oxidation of the double bond were some of the postfunctionalization reactions performed in the newly synthetized cycloadducts that demonstrated the synthetic utility of the protocol *(40)*.

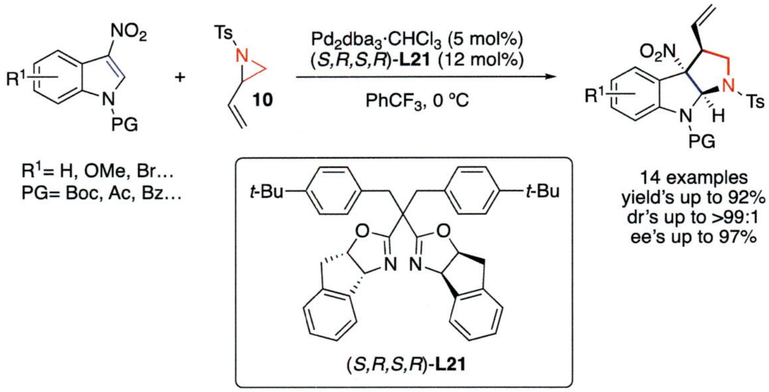

Scheme 29 Dearomative Pd-catalyzed [3 + 2] cycloaddition reaction between vinyl aziridine **10** and 3-nitroindoles to yield 3a-aminopyrroloindolines.

In a similar fashion, Hou and coworkers reported the Pd-catalyzed [3 + 2] cycloaddition reaction between vinyl aziridine **10** and 3-nitroindoles *(41)*. The use of WalPhos-type ligand (S,S_P)-**L22** provided high diastereo- and enantioselectivities for a range of 3-nitroindoles (up to >20:1 dr and up to 95% ee; Scheme 30A). Moreover, substituting the nitrogen atom of the indole for a sulfur also gave access to thiophene–fused pyrrolidines in good diastereo- and enantioselectivity (4:1 dr and 71% ee; Scheme 30B).

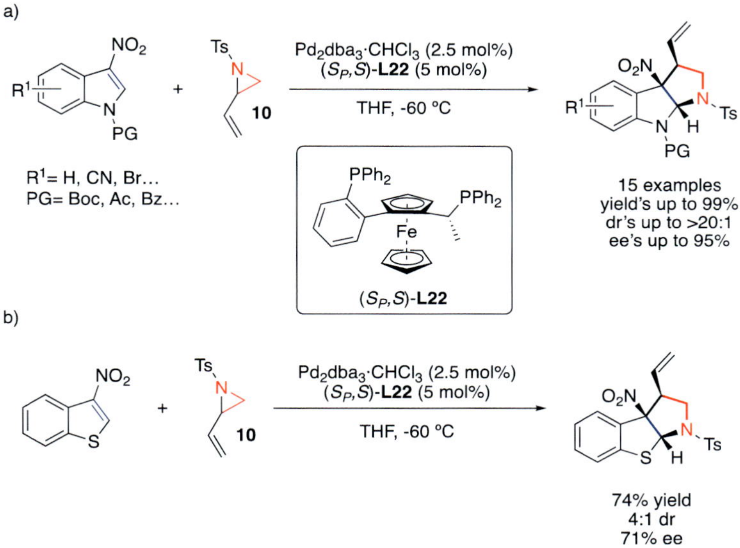

Scheme 30 Pd/(S,Sp)-**L22** [3+2] cycloaddition reaction of vinyl aziridine **10** with (A) 3-nitroindoles and (B) 3-nitrobenzo[b]thiophene.

In 2020, Enders and coworkers found out an efficient synthesis of indanone-fused spiropyrrolidines via Pd-catalyzed [3+2] cycloaddition of vinyl aziridine **10** with a range of indanone 1,3-diones *(42)*. Surprisingly, after ligand screening, only phosphine ligands with a pendant Boc-protected amine group (such as ligand (S,S)-**L23**) yielded the cycloadducts in a stereoselective manner (up to 10:1 dr and up to 94% ee; Scheme 31). The authors propose that the amide moiety in the ligand participates in the activation of the dienone by means of hydrogen bonding (Scheme 31).

Another approach to obtain highly functionalized pyrrolidines is through a Pd-catalyzed [3+2] cycloaddition reaction of activated vinyl cyclopropanes with imines. In this context, the groups of Guo and Vitale simultaneously reported in 2018 a Pd-catalyzed [3+2] cycloaddition of nitrile-activated vinyl cyclopropane **3** with benzo[e][1,2,3]oxathiazine 2,2-dioxide *(43,44)*. Even though, both groups achieved good yields and diastereoselectivities using ligands (R)-**L24** (Guo's group) and (S,S)-**L25** (Vitale's group), low ee's were attained (up to 22% ee; Scheme 32).

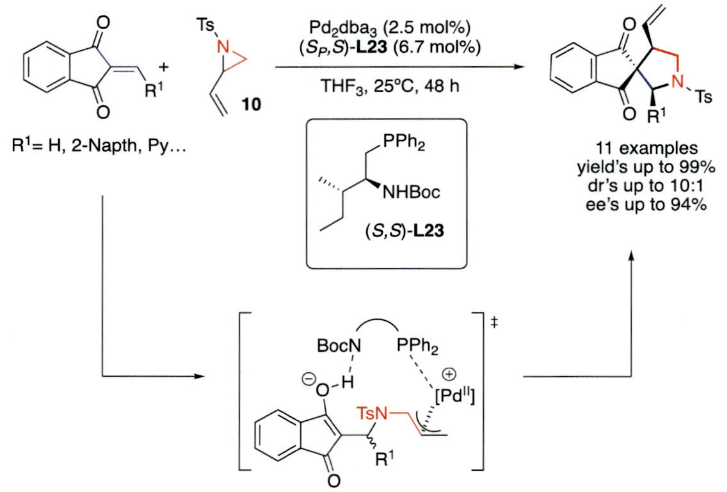

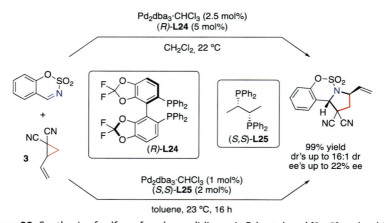

Scheme 31 Indanone-fused spiropyrrolidines obtained via Pd-catalyzed [3 + 2] cycloaddition of vinyl aziridine **10** with indane-1,3-diones.

Scheme 32 Synthesis of sulfone-fused pyrrolidines via Pd-catalyzed [3 + 2] cycloaddition of nitrile-activated vinyl cyclopropane **3** with benzo[e][1,2,3]oxathiazine 2,2-dioxide.

Later, Liu and coworkers developed a Pd-catalyzed [3 + 2] cycloaddition of activated vinyl cyclopropanes **11** and **12** with N-isatin derivatives and aldimines *(45)*. In the case of isatin derivatives phosphoroamidite ligand (S)-**L26** provided the best diastereo- and enantioselectivities (up to 5:1 dr and up to 96% ee; Scheme 33A), whereas to achieve high selectivities for aldimines the use of ligand (R,R,R)-**L27** was necessary (up to 7:1 dr and up to 93% ee; Scheme 33B). The synthetic potential of the catalytic protocol

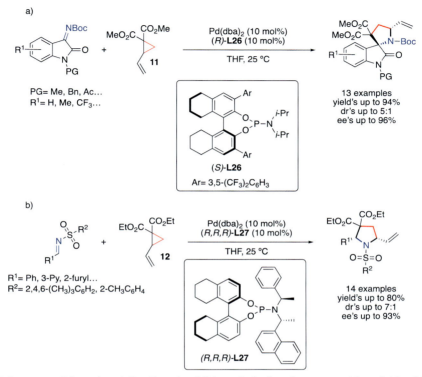

Scheme 33 Pd-catalyzed [3+2] cycloaddition of vinyl cyclopropane **11** and **12** with (A) *N*-isatin derivatives and (B) aldimines.

was further proved by the post-functionalization of the newly formed cycloadducts *(45)*.

In 2019, Ni and coworkers reported one example of Pd-catalyzed [3+2] cycloaddition of the activated vinyl cyclopropane **11** to (*E*)-*N*-(benzo[*d*]thiazol-2-yl)-1-phenylmethanimine *(46)*. The use of (*S*)-BINAP yielded the corresponding pyrrolidine in moderate diastereoselectivity and high ee (up to 94% ee; Scheme 34).

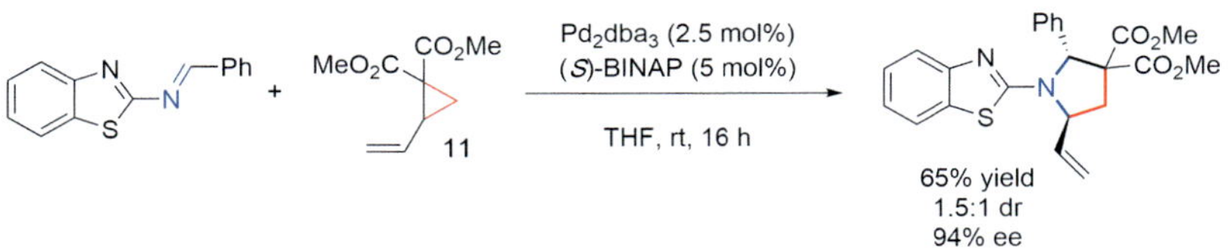

Scheme 34 Pd-catalyzed [3+2] cycloaddition of the activated vinyl cyclopropane **11** to (*E*)-*N*-(benzo[*d*]thiazol-2-yl)-1-phenylmethanimine.

Pyrrolidines can also be obtained via Pd-TMM [3+2] cycloaddition reactions. For instance, Trost and coworkers reported in 2020 a successful example of Pd-catalyzed [3+2] cycloaddition between a phosphonate-methylallyl **13** with the electrodeficient aldimine **14** (>20:1 dr and 98% ee; Scheme 35) *(47)*.

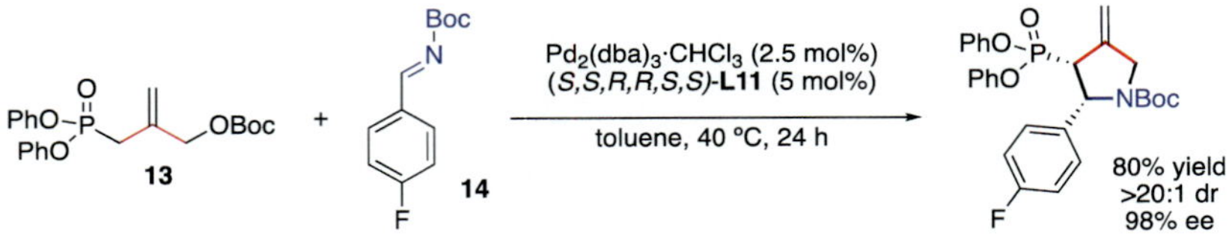

Scheme 35 Pd/(*S,S,R,R,S,S*)-**L11** [3+2] cycloaddition of phosphonate-methylallyl **13** with electrodeficient imine **14**.

Later, the same group extended the Pd–TMM cycloaddition methodology to the use of fluorinated methylallyl donors with a range of aldimines *(21)*. Thus, a range of fluorinated containing pyrrolidines were synthetized in high diastereo- and enantioselectivities by using diamidophosphite ligand (*S,S,R,R,S,S*)-**L11** (up to >50:1 dr and up to 99% ee; Scheme 36).

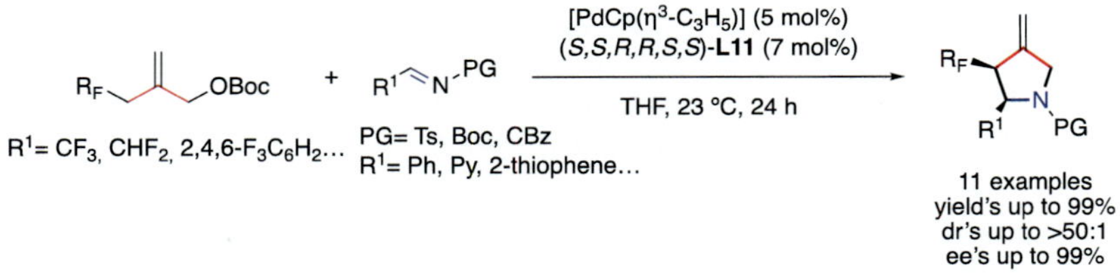

Scheme 36 Pd catalyzed [3+2] cycloaddition reaction of CF_3-methylallyl donors with imines.

3.2 Indoles

A large number of pharmaceuticals present in their core structure a fused azabicyclic compounds and indole-related structures. In recent years, these compounds are generally synthetized using the Pd-catalyzed [3+2] cyclo-addition reaction *(48)*. For instance, Li and coworkers reported an unusual approach to obtain indolines via the Pd-catalyzed ring contraction of vinyl benzoxazinanones *(49)*. The key step in this protocol is the intramolecular nitrogen nucleophilic attack to C2, leading to the palladacyclobutane inter-mediate, which upon β-hydride elimination provided the desired indoles (Scheme 37). Good enantioselectivities were attained using chiral diphosphine (*R,R*)-**L28** (up to 89% ee).

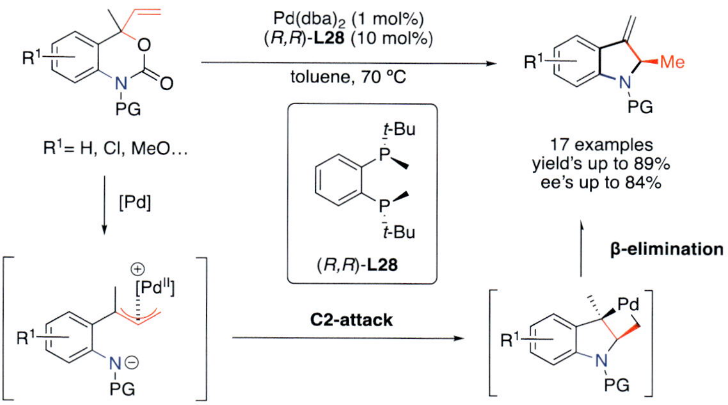

Scheme 37 Synthesis of chiral indolines via the Pd-catalyzed ring contraction of vinyl benzoxazinanones.

3.3 Imidazolidines

Imidazolines that present an aminal (*N*,*N*-acetal) scaffold are present in many natural products, bioactive molecules and ligands *(50)*. In 2018, Campange and coworkers reported in two sequential papers on a convenient way to synthetize *N*-sulfoximides fused with imidazolidines via the Pd-catalyzed [3 + 2] cycloaddition of *N*-sulfoximides and substituted vinyl aziridines *(51,52)*. High yields, diastereo- and enantioselectivities were achieved using ligand (*R*)-**L29** (up to >20:1 dr and up to 90% ee; Scheme 38).

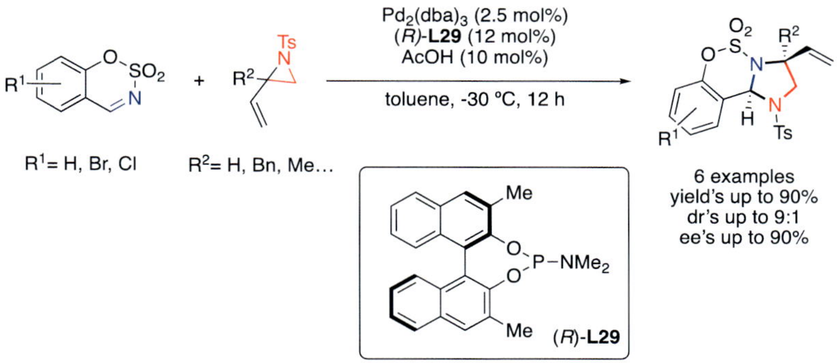

Scheme 38 Pd/(*R*)-**L29** catalyzed [3 + 2] cycloaddition of *N*-sulfoximides with substituted vinyl aziridines.

3.4 Tetrahydroquinolines

Highly decorated tetrahydroquinolines represent the core of many molecules with physiological activity *(53)*. One of the most interesting strategies to build this complex molecules is trough Pd-catalyzed [4 + 2] cycloaddition reaction of vinyl benzoxazinanones with electrodeficient alkenes *(54)*. Thus, Shi and coworkers reported in 2017 a facile strategy to build spiroxindole fused tetrahydroquinolines via the Pd-catalyzed [4 + 2] cycloaddition of vinyl benzoxazinanones with methylene indolinones *(55)*. After ligand optimization it was found that spirophosphine ligand (*S*)-**L30** provided the best diastereo- and enantioselectivities (up to >20:1 dr and up to 99% ee; Scheme 39).

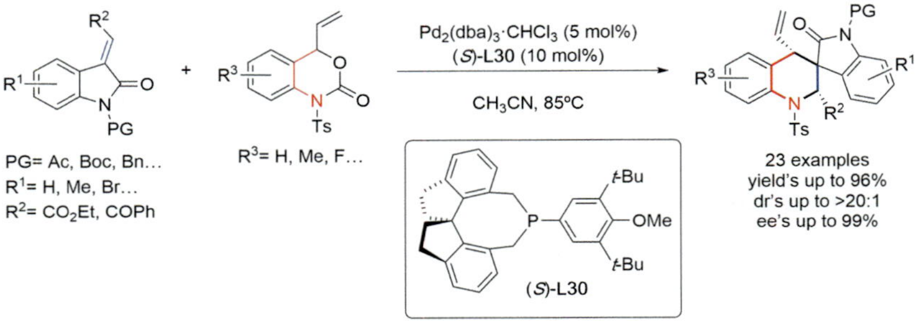

Scheme 39 Synthesis of spiroxindole fused tetrahydroquinolines via the Pd-catalyzed [4 + 2] cycloaddition of vinyl benzoxazinanones with methylene indolinones.

In 2018, Song and coworkers developed an interesting way to synthetize barbiture-fused spirotetrahydroquinolines using the Pd-catalyzed [4 + 2] cycloaddition approach by reacting vinyl benzoxazinanones with barbiture-derived olefins *(56)*. The use of Pd/(*R*)-**L31** catalytic system yielded high diastereo- and enantioselectivities (up to >99:1 dr and up to 97% ee; Scheme 40).

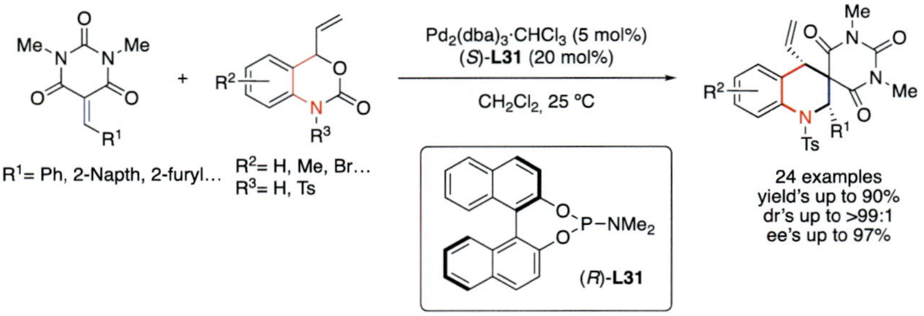

Scheme 40 Pd-catalyzed [4 + 2] cycloaddition of vinyl benzoxazinanones with barbiture-derived olefins.

Later, Hou and coworkers developed a Pd–catalyzed [4 + 2] cycloaddition of vinyl benzoxazinanones with 3-nitroindoles to synthetize highly decorated nitro–indoline scaffolds *(57)*. Ferrocenyl ligand (R,S,S_P)-**L32** yielded the best diastereo- and enantioselectivities (up to >20:1 dr and up to 95% ee; Scheme 41).

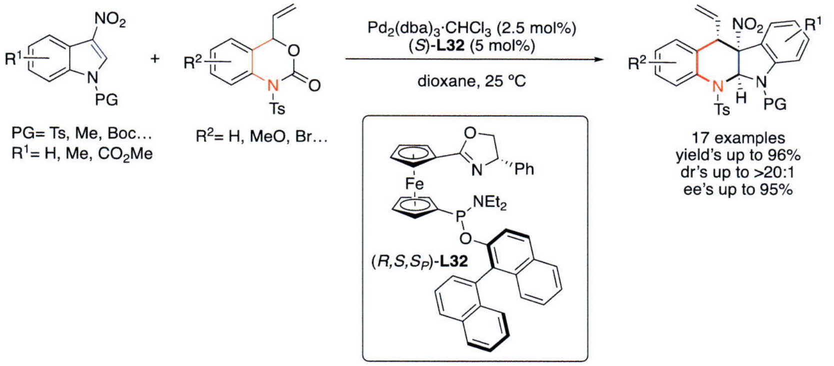

Scheme 41 Synthesis nitro-indoline scaffolds via Pd-catalyzed [4 + 2] cycloaddition of vinyl benzoxazinanones with 3-nitroindoles.

Very recently, Zhong and coworkers reported the [4 + 2] cycloaddition reaction of vinyl benzoxazinanones with alkylidene pyrazolones to synthetize spiropyrazolones fused with a tetrahydroquinoline scaffold *(58)*. Ligand screening disclosed phosphoroamidite ligand (S,S,S)-**L9** as the optimal choice to attain high diastereo- and enantioselectivities (up to >99:1 dr and up to 99% ee; Scheme 42). Gram–scale synthesis was performed to prove the synthetic utility of the protocol *(58)*.

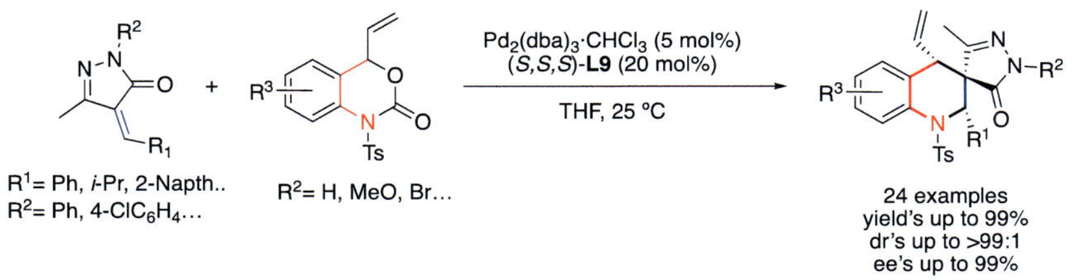

Scheme 42 Pd-catalyzed [4 + 2] cycloaddition reaction of vinyl benzoxazinanones with alkylidene pyrazolones.

3.5 Tetrahydroquinazolines

Some antibacterial, antidepressive, anti–inflammatory and broncho ventilator pharmaceuticals contain tetrahydroquinazolines scaffolds in their core structure *(59)*. Guo and coworkers developed a facile way to synthetize sulfamate-fused tetrahydroquinazolines via Pd-catalyzed [4 + 2] cycloaddition of vinyl benzoxazinanones and sulfamate-derived imines *(60)*. Pd/(*R*)-**L2** catalytic system yielded the best diastereo- and enantioselectivities in a broad range of vinyl benzoxazinanones as well as of imines (up to >20:1 dr and up to 96% ee; Scheme 43). Several transformations were performed in the newly formed cycloadducts such as: reduction of the vinyl moiety, tandem hydroboration-oxidation and bromation reactions *(60)*.

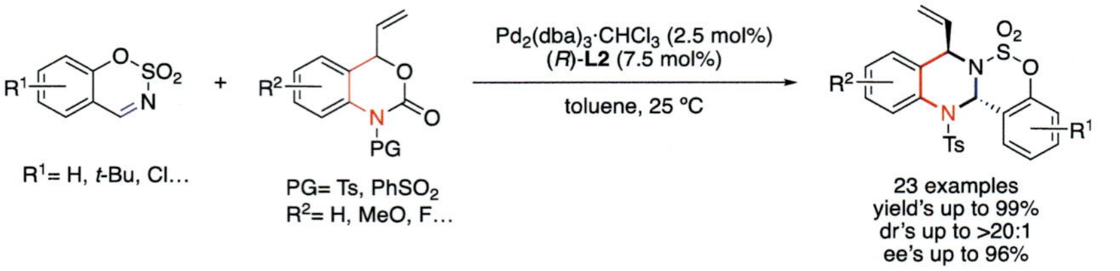

Scheme 43 Pd-catalyzed [4+2] cycloaddition of vinyl benzoxazinanones and sulfamate-derived imines.

In 2019, Kim and coworkers developed a very similar [4 + 2] cycloaddition of vinyl benzoxazinanones and sulfamate-derived imines to synthetize sulfamate-fused tetrahydroquinazolines *(61)*. An although the authors report the formation of the same cycloadducts a slight modification in the reaction conditions as well as on the configuration of the ligand to (*S*)-**L2** provided the inverse enantiomer in high selectivities (up to >30:1 dr and up to 98% ee; Scheme 44).

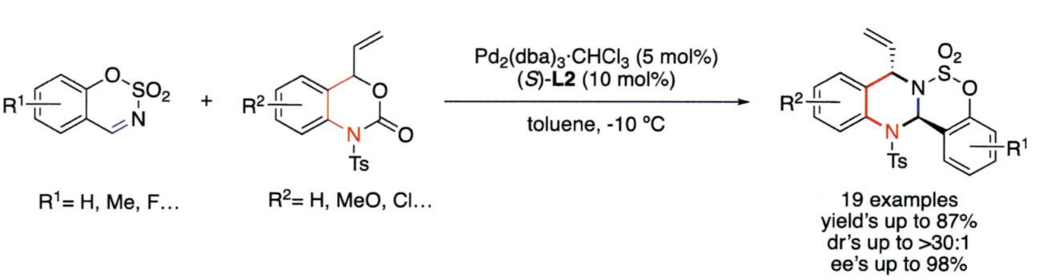

Scheme 44 Pd-catalyzed [4+2] cycloaddition of vinyl benzoxazinanones and sulfamate-derived imines to synthetize sulfamate-fused tetrahydroquinazolines.

3.6 Dihydroquinazolines

Dihydroquinazoline scaffolds are part of a privileged family of bioactive natural alkaloids known as tryptanthrins *(62)*. In 2017, Shi and coworkers managed to synthetize the dihydroquinazoline core of these molecules via the decarboxylative Pd-catalyzed [4 + 2] cycloaddition of benzoxazinanones with isatins *(63)*. After optimization, the use of spirophosphine (*S*)-**L30** yielded the best enantioselectivities for a range of both benzoxazinanones and of isatin derivatives (up to >99% ee; Scheme 45).

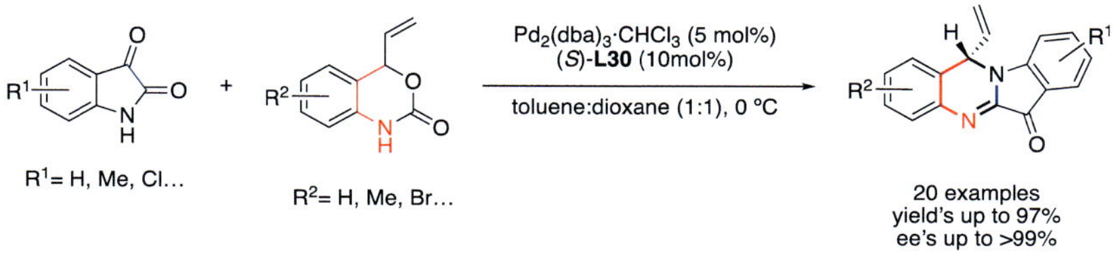

Scheme 45 Pd/(*S*)-**L30** catalyzed [4 + 2] cycloaddition of benzoxazinanones with isatins.

3.7 Pyrazolo hexahydroisoquinolines

Pyrazolo[5,1,*a*]isoquinoline derivatives form a complex group of cyclo-adducts that has been proven to have several uses as bioactive molecules *(64)*. In 2019, Guo and coworkers reported a facile pathway to synthetize complex hexahydro pyrazolo[5,1,*a*]isoquinolines scaffolds via the tandem Pd-TMM [3 + 2] cycloaddition/allylation of trimethylsilyl–allenyl acetates with azomethine imines *(65)*. Several cycloadducts could be synthetized in high enantioselectivity and high *E/Z* ratio with the use of phospho-roamidite ligand (*S*)–**L2** (up to 5:1 *E/Z* and up to 99% ee; Scheme 46A).

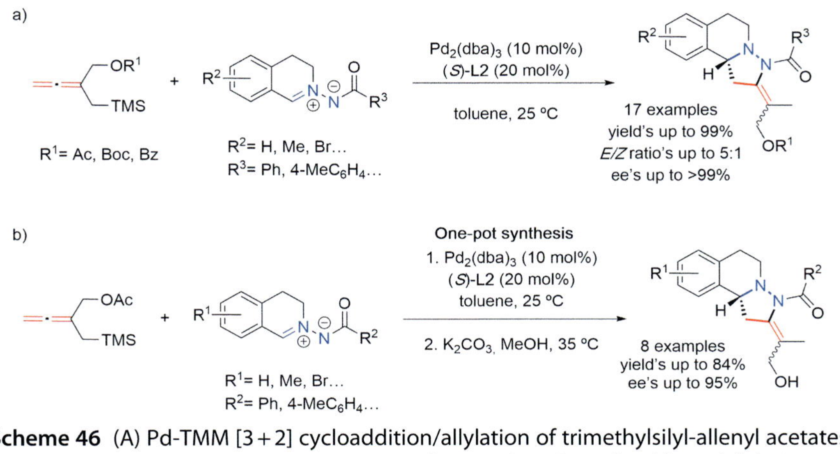

Scheme 46 (A) Pd-TMM [3 + 2] cycloaddition/allylation of trimethylsilyl-allenyl acetates with azomethine imines. (B) One-pot Pd-TMM [3 + 2] cycloaddition/allylation of 2-((trimethylsilyl)methyl)buta-2,3-dien-1-yl acetate with azomethine imines and saponification of the acetate.

Moreover, the one pot synthesis of allyl alcohols could be performed without significant erosion of the enantioselectivity (up to 95% ee; Scheme 46B).

Later that year, the same group reported the synthesis of tetrahydro pyrazolo[5,1,*a*]isoquinoline derivatives via de Pd-catalyzed [3 + 2] annulation of 1-phenylpropa-1,2-dien-1-yl acetate and azomethine imine **15** *(66)*. Although high yields could be obtained only moderate enantioselectivities (up to 80% ee) were attained using phosphoroamidite ligand (*S*)-**L2** (Scheme 47).

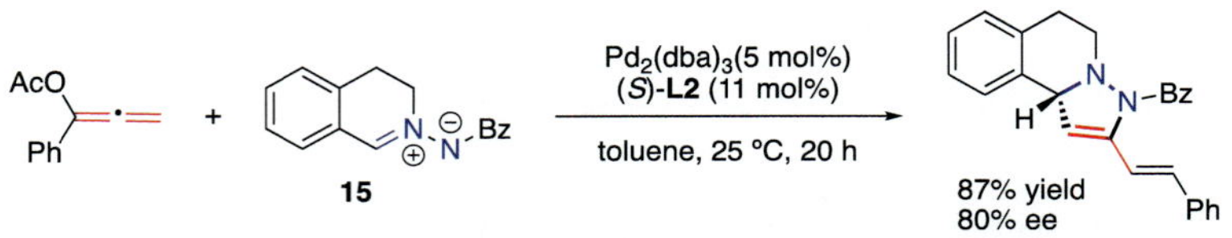

Scheme 47 Pd-catalyzed [3 + 2] cyclization of 1-phenylpropa-1,2-dien-1-yl acetate and azomethine imine **15**.

3.8 Azepines

Seven-membered aza-heterocycles, among them azepines, represent a privileged structure embedded in many pharmaceutical and natural products *(67)*. In 2019, Chen an coworkers developed a dual cooperative tertiary amine/iridium catalyzed cycloaddition between isatin derived Morita-Baylis–Hillman (MBH) carbonates and carbamate-functionalized allyl carbonates to yield the seven-membered cycloadducts selectively *(68)*. They concluded that the combination of the phosphoroamidite containing Ir-complex **C1** with DABCO as a base provided the best diastereo- and enantioselectivities (up to >19:1 dr and up to 95% ee; Scheme 48A). They also found that carbamate-functionalized allyl carbonate substrate could be efficiently substituted by methyl 2-oxo-6-vinyl-1,3-oxazinane-3-carboxylate attaining high diastereo- and enantioselectivities using Ir-complex **C2** and **16** as a base (up to >19:1 dr and up to 92% ee; Scheme 48B).

Next year, Deng and coworkers used the Pd-TMM approach to synthetize benzofuran-fused azepines via [4 + 3] cycloaddition of imino-methylallyl substrate **17** to azadiene substrates *(69)*. By using phosphoroamidite ligand (*S,S,S*)-**L9** high diastereo- and enantioselectivities were attained (up to >20:1 dr and up to 99% ee; Scheme 49A). Moreover, high diastereo- and enantioselectivities were also attained when using substrate **4** with phosphoroamidite ligand (*S*)-**L33** (up to >20:1 dr and up to >99% ee; Scheme 49B). Interestingly, this modification of the electron

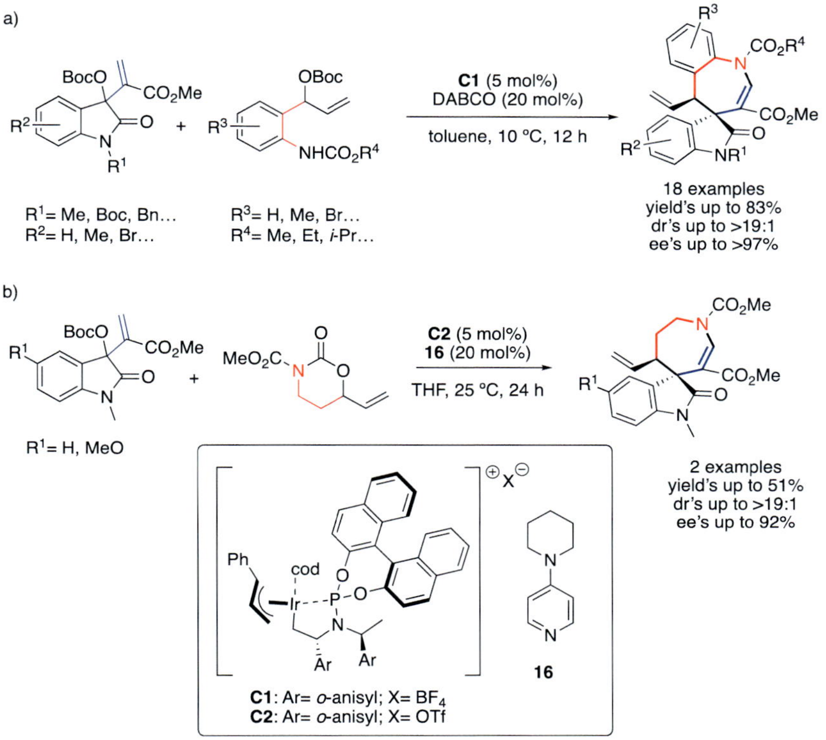

Metal-π-allyl mediated asymmetric cycloaddition reactions 137

a)

R^1= Me, Boc, Bn...
R^2= H, Me, Br...

R^3= H, Me, Br...
R^4= Me, Et, i-Pr...

C1 (5 mol%)
DABCO (20 mol%)

toluene, 10 °C, 12 h

18 examples
yield's up to 83%
dr's up to >19:1
ee's up to >97%

b)

R^1= H, MeO

C2 (5 mol%)
16 (20 mol%)

THF, 25 °C, 24 h

2 examples
yield's up to 51%
dr's up to >19:1
ee's up to 92%

C1: Ar= o-anisyl; X= BF$_4$
C2: Ar= o-anisyl; X= OTf

Scheme 48 Dual cooperative tertiary amine/iridium catalyzed [4+3] cycloaddition between isatin derived Morita-Baylis-Hillman (MBH) carbonates and (A) carbamate-functionalized allyl carbonates and (B) cyclic vinyl carbamate.

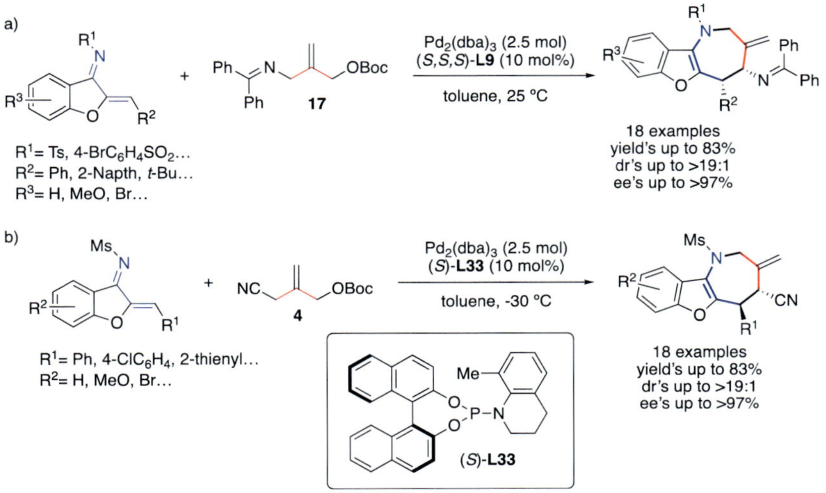

a)

R^1= Ts, 4-BrC$_6$H$_4$SO$_2$...
R^2= Ph, 2-Napth, t-Bu...
R^3= H, MeO, Br...

Pd$_2$(dba)$_3$ (2.5 mol)
(S,S,S)-L9 (10 mol%)

toluene, 25 °C

18 examples
yield's up to 83%
dr's up to >19:1
ee's up to >97%

b)

R^1= Ph, 4-ClC$_6$H$_4$, 2-thienyl...
R^2= H, MeO, Br...

Pd$_2$(dba)$_3$ (2.5 mol)
(S)-L33 (10 mol%)

toluene, -30 °C

18 examples
yield's up to 83%
dr's up to >19:1
ee's up to >97%

(S)-L33

Scheme 49 Pd-catalyzed [4+3] cycloaddition of azadienes to (A) imino-methylallyl substrate **17** and (B) cyano-methylallyl substrate 4.

withdrawing substituent on the TMM substrate causes a shift of the dia-stereoselectivity achieving the trans adduct preferentially (Scheme 49B).

At the same time, Zuo and coworkers reported the synthesis a series of the same benzofuran N-seven-membered rings via the Pd–TMM catalyzed [4 + 3] cycloaddition of imino-methylallyl substrates to a series of azadiene compounds (35). The use of phosphoramidite ligand (R,S,S)-**L9** yielded the trans cycloadducts in high diastereo- and enantioselectivities (up to >19:1 dr and up to 98% ee; Scheme 50).

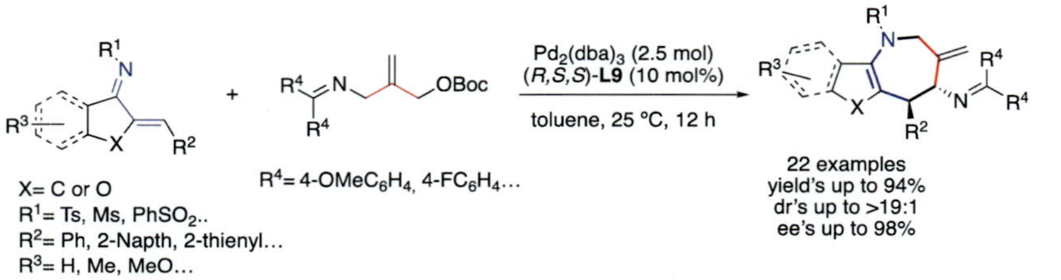

Scheme 50 Pd-TMM catalyzed [4 + 3] cycloaddition of imino-methylallyl substrates with a series of azadiene compounds.

The same year, Shao and coworkers reported the similar Pd–catalyzed [4 +3] cycloaddition of imino-methylallyl substrate **17** with a range of azadiene compounds to synthetize benzofuran–fused azepines (70). Octahydro–BINOL based phosphoramidite ligand (R,R,R)–**L34** yielded high diastereo- and enantioselectivities for several azadienes (up to >19:1 dr and up to 99% ee; Scheme 51).

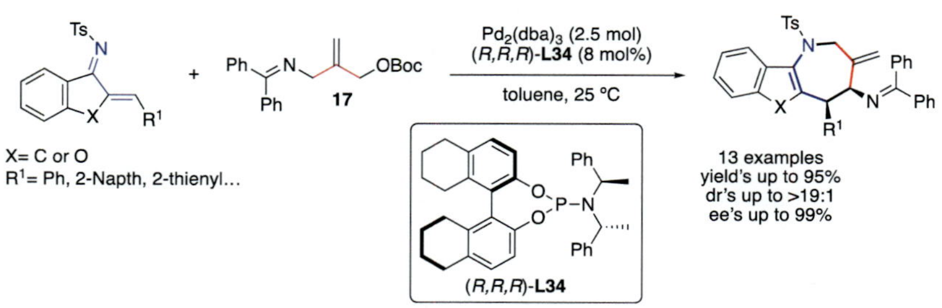

Scheme 51 Pd-catalyzed [4 + 3] cycloaddition of imino-methylallyl substrate **17** with azadiene derivatives.

Following the lead of previous reports, Deng and coworkers developed a synthetic pathway to obtain indole-fused azepines using the Pd–catalyzed [4 + 3] cycloaddition of cyano-methylallyl substrate **4** with a series of indole–derived azadiene substrates *(71)*. The cycloadducts were attained in high diastereo– and enantioselectivities using phosphoroamidite ligand (*S*)-**L35** (up to >20:1 dr and up to 99% ee; Scheme 52).

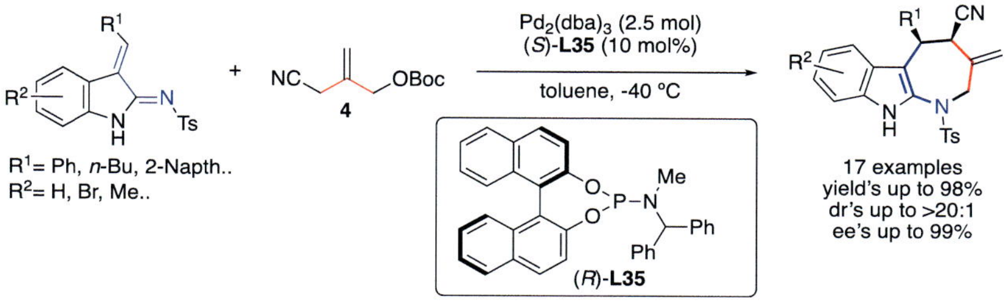

Scheme 52 Pd-catalyzed [4+3] cycloaddition of cyano-methylallyl substrate **4** to a series of indole-derived azadiene substrates.

In 2021, the same group developed a similar way to synthetize benzofuran–fused azepines using the Pd–TMM catalyzed [4 + 3] cycloaddition approach but in this case with a transient *p*-nitrophenylsulfonyl (Ns) group attached to methylallyl substrate **18** *(72)*. Phosphoroamidite ligand (*S*)-**L36** was used to afford the sulphonated azepines in high diastereo– and enantioselectivities (up to 20:1 dr and up to 99% ee; Scheme 53A). Moreover, the sulphonyl group could be removed easily and with exclusive regioselectivity maintaining high yields and enantioselectivities using racemic Pd–catalyst (up to 99% ee; Scheme 53B).

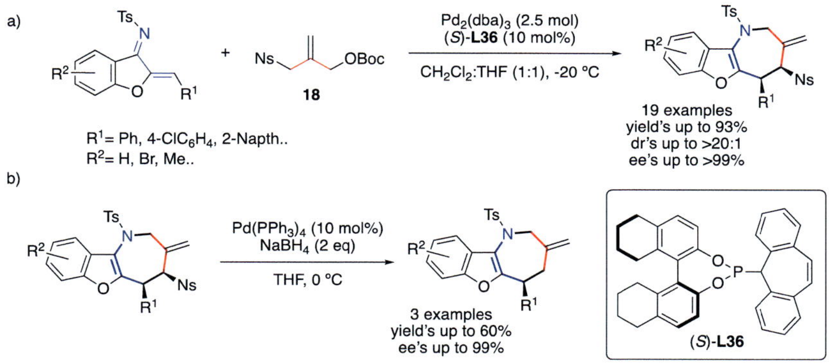

Scheme 53 (A) Pd-TMM catalyzed [4+3] cycloaddition of azadienes with a transient sulphonate-methylallyl substrate **18**. (B) Catalytic removal of the Ns group.

3.9 Lactams

Chiral medium sized lactams are part of many natural recurring motifs as well as important intermediates for the synthesis of chiral synthons *(28,73)*. In 2017, Glorius and coworkers investigated the mechanistic insights of a previously reported Pd-catalyzed [4 + 3] annulation of vinyl benzoxazinanone **19** with cinnamaldehyde to yield chiral benzazepines *(3a,3b)*. The reaction was monitored via ESI–MS, in situ NMR studies and the structures of key complexes were determined via X-Ray diffraction. All these techniques led to the conclusion that the high diastereo selectivity and enantioinduction of the reaction was due to the formation of a Pd/(R,S)-**L37**/PPh$_3$ homoenolate complex (>20:1 dr and 99% ee; Scheme 54).

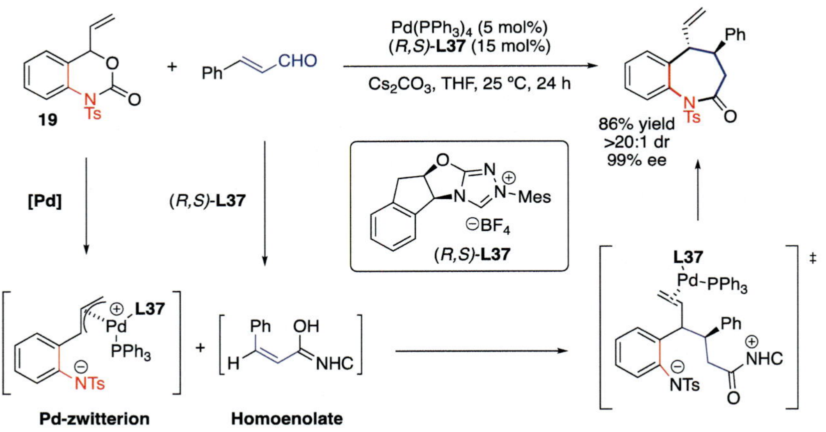

Scheme 54 Pd-catalyzed [4+3] annulation of vinyl benzoxazinanone **19** with cinnamaldehyde.

Later the same year, Xiao and coworkers developed an efficient synthesis benzoquinol-2-ones via Pd-catalyzed [4 + 3] cycloaddition of vinyl benzoxazinanones with *in situ* photochemically generated ketenes from α-diazoketones *(3c)*. The best results were attained using P,S ligand (S,R)-**L10** obtaining high diastereo- and enantioselectivities (up to >20:1 dr and up to 96% ee; Scheme 55). Post functionalization reaction could be performed on the newly synthetized cycloadducts. Moreover, the synthetic utility of the methodology was further demonstrated with the gram-scale transformation in flow with sunlight as photochemical source and low catalyst loadings *(3c)*.

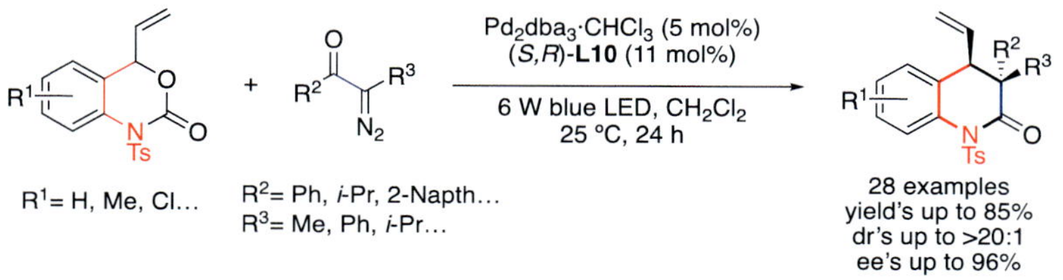

Scheme 55 Pd-catalyzed [4+3] cycloaddition of vinyl benzoxazinanones with in situ photochemically generated ketenes from α-diazoketones.

Next year, Deng and coworkers reported the decarboxylative Pd-catalyzed [4+2] cycloaddition of vinyl benzoxazinanones with carboxylic acids to synthetize 3,4-dihydroquinolin-2-ones *(74)*. After optimization chiral phosphine (R)-**L38** was selected as the ligand of choice affording high diastereo- and enantioselectivities (up to >99:1 dr and up to 97% ee; Scheme 56).

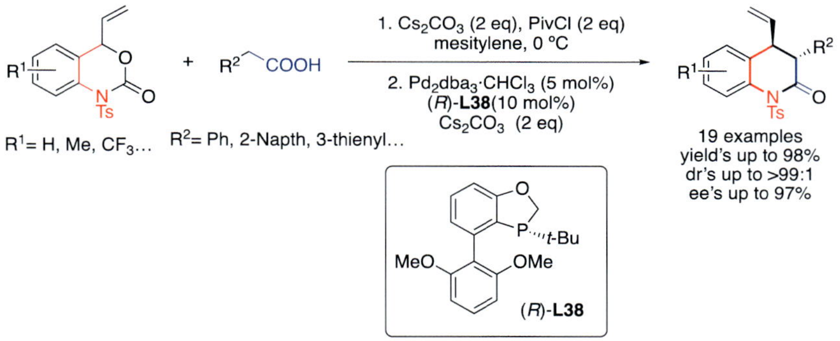

Scheme 56 Pd-catalyzed [4+2] cycloaddition of vinyl benzoxazinanones with carboxylic acids.

In 2019, Xiao and coworkers developed an asymmetric Pd-catalyzed [4+2] cycloaddition reaction of vinyl benzoxazinanones with butenolides to efficiently synthetize dihydroquinol-2-ones *(75)*. Thioether–phosphoro-amidite ligand (R,S,R)-**L39** enabled this transformation in high diastereo- and enantioselectivities (up to >20:1 dr and up to 95% ee; Scheme 57A). DFT calculations suggest that this high selectivity values are due to combination of the steric hindrance induced by the chiral ligand in addition to the formation of hydrogen bonds between the substrates achieving the optimal orientation *(75)*. Interestingly this reaction could also be carried out using

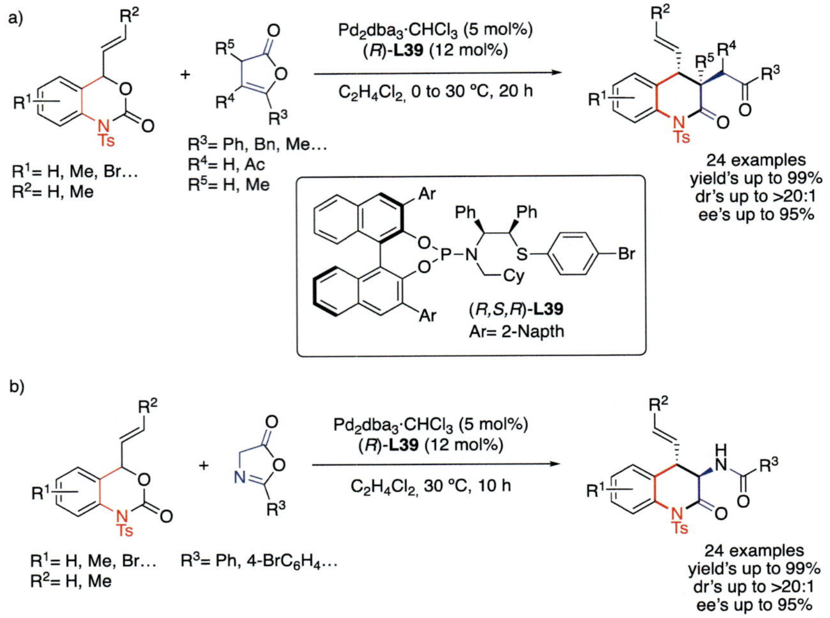

Scheme 57 Pd-catalyzed asymmetric [4+2] cycloaddition reaction of vinyl benzoxazinanones with (A) butenolides and (B) azalactones.

azalactones instead of butenolides under the same reaction conditions reported for the latter, yielding the corresponding amido substituted quinol-2-ones with high selectivities (up to >20:1 dr and up to 92% ee; Scheme 57B).

Later, Du and coworkers developed a synergistic Pd/NHC-catalyzed [3+2] cycloaddition reaction of vinyl aziridines with 3-substituted but-2-enoates yielding vinylpyrrolidin-2-ones *(76)*. Although good yields could be obtained through this cooperative Pd/NHC approach, the attempts of performing an asymmetric version of this reaction provided modest enantioselectivities when using carbene precursor **20** in combination with chiral squaramide **21** (up to 65% ee; Scheme 58).

Yang, Deng and coworkers later disclosed an innovative pathway to form chiral indole-fused 2-pyrrolidones through a dual Ir/Lewis base-catalyzed [3+2] cycloaddition of vinyl indoloxazolidones with carbonylic anhydrides *(77)*. The combination of Ir-complex **C3** and base **23** resulted in the best diastereo- and enantioselectivities for a range of substrates (up to >20:1 dr and up to 99% ee; Scheme 59).

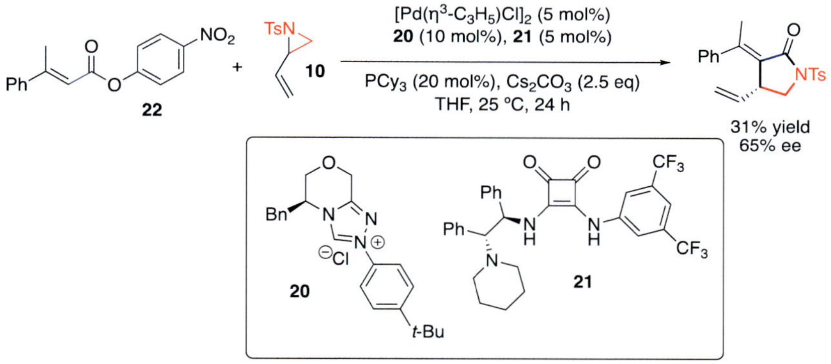

Scheme 58 Synergistic Pd/NHC-catalyzed [3+2] cycloaddition reaction of vinyl aziridine **10** with 3-substituted but-2-enoate **22**.

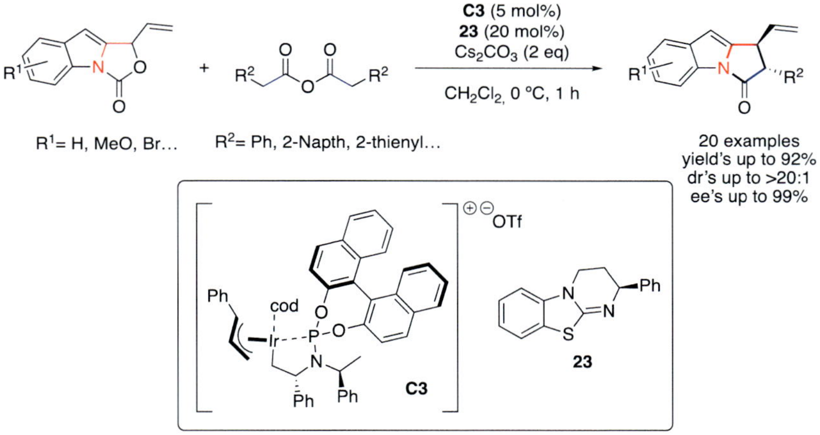

Scheme 59 Preparation of chiral indole-fused 2-pyrrolidones through a dual Ir/Lewis base-catalyzed [3+2] cycloaddition of vinyl indoloxazolidones with carbonylic anhydrides.

In 2021, Glorius and coworkers reported the synthesis of *N*-benzoyl-1,2,-hydroquinolines via the [4+2] cycloaddition reaction of vinyl benzoxazinanones with 2-chloro-3-phenylpropanal catalyzed by a cooperative Ir/NHC system *(30)*. They disclosed that by combining phosphoroamidite ligand (*S*)-**L16** and the NHC precursor (*R*,*S*)-**L37**-(Cl) high diastereo- and enantioselectivities were attained (up to >20:1 dr and up to 99% ee; Scheme 60).

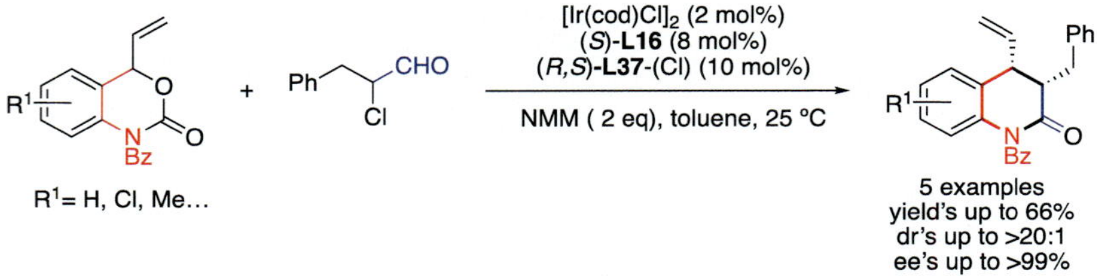

Scheme 60 Dual Ir/NHC catalyzed [4+2] cycloaddition reaction of vinyl benzoxazinanones with 2-chloro-3-phenylpropanal.

3.10 Ureas

Urea–related cyclic compounds are present in many biologically relevant structures as well as in some useful chiral building blocks *(78)*. In this context, in 2018 Shi and coworkers synthesized dihydroquinazolinone scaffolds via the Pd-catalyzed [4+2] cycloaddition of vinyl benzoxazinanones with isocyanates *(79)*. The use of phosphoramidite ligand (*S*)-**L40** yielded the cycloadducts in high enantioselectivities for a series of vinyl benzoxazinanones and of isocyanates (up to 92% ee; Scheme 61).

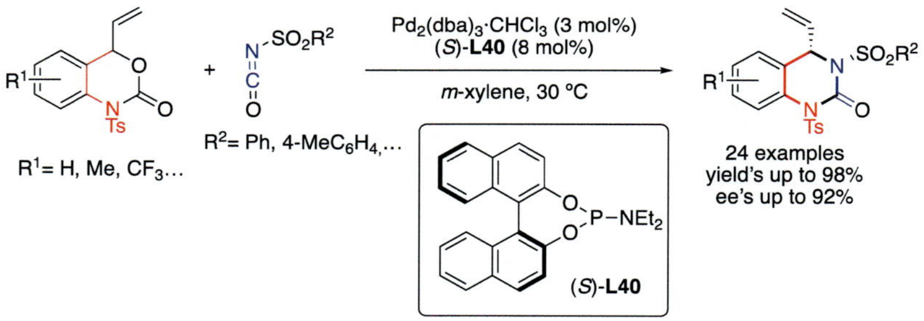

Scheme 61 Pd-catalyzed [4+2] cycloaddition of vinyl benzoxazinanones with isocyanates.

Next year, Zhang and coworkers developed an interesting way to synthetize cyclic ureas via the Pd-catalyzed [3+2] cycloaddition of *N*-containing allylic carbonates **24** and **25** and isocyanates *(80)*. The use of chiral phosphoramidite (*S,R,R*)-**L9** provided the optimal enantioselectivities for the synthesis of five membered rings (up to 98% ee; Scheme 62A), while for six-membered ring ureas the best enantioselectivities were attained using phosphoramidite (*R*)-**L2** (up to 99% ee; Scheme 62B).

In 2021, Shi and coworkers disclosed the synthesis of indole-fused imidazolin-2-one derivate via the decarboxylative Pd-catalyzed [3+2] annulation of indole-based carbamate **26** with 4-methylbenzenesulfonyl

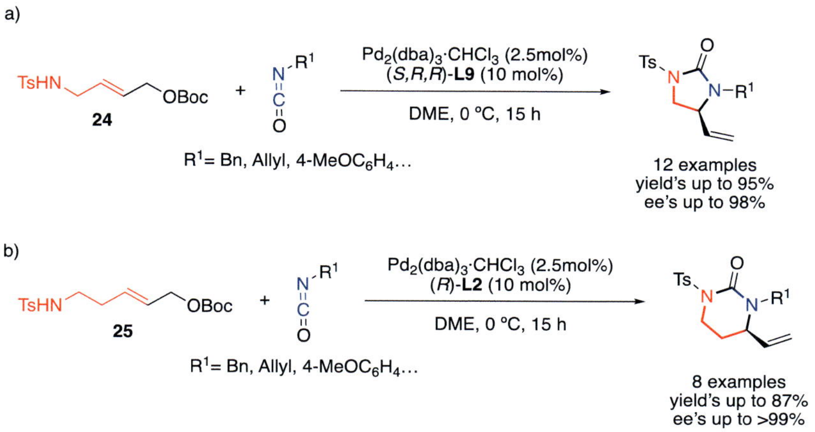

Scheme 62 Pd-catalyzed [3+2] cycloaddition isocyanates to N-containing allylic carbonates (A) **24** and (B) **25**.

isocyanate *(81)*. Even though high yields could be obtained the development of an asymmetric version remains a challenge since low yields and enantioselectivies were achieved with spirophosphoroamidite ligand (R,R,R)-**L41** (up to 20% ee; Scheme 63).

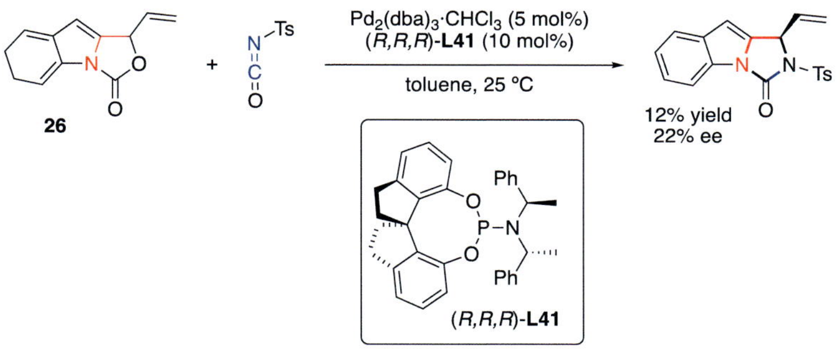

Scheme 63 Pd-catalyzed [3+2] annulation of indole-based carbamate **26** with methylbenzenesulfonyl isocyanate.

4. Cycloaddition reactions for the formation of carbocycles

4.1 Cyclopentanes

Cyclopentane moieties are commonly found in a wide variety of natural products *(82)*. A powerful approach to obtain this kind of scaffolds is through Pd-catalyzed asymmetric [3 + 2] cycloadditions of a 1,3-dipole generated by the opening of vinyl cyclopropanes (VCPs) or other strained rings with electron deficient olefins *(83)*. For instance, Wang and coworkers reported in 2018 a methodology for the synthesis of indole-fused cyclopentanes involving the Pd-catalyzed dearomative [3+2] cycloaddition of VCP **27** to nitroindoles *(40)*. After ligand screening, the use of bisoxazoline ligand (S,R,S,R)-**L42** was found to be optimal, leading to moderate diastereoselectivities and high enantioselectivities for a range of indoles with different substitution pattern and electronic properties in the aryl ring (up to 3:1 dr and up to 92% ee; Scheme 64).

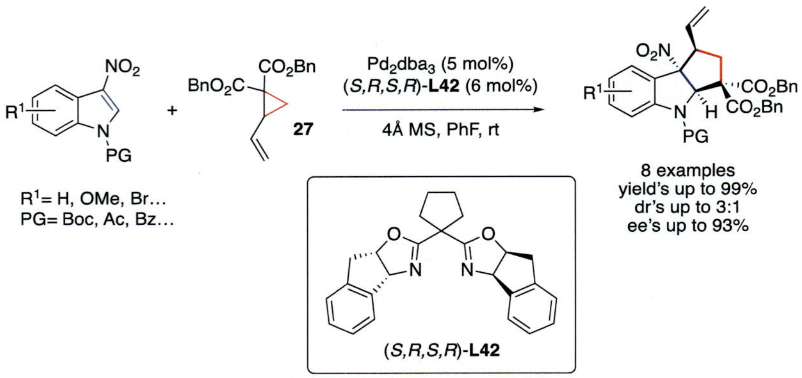

Scheme 64 Synthesis of indole-fused cyclopentanes via Pd/(S,R,S,R)-**L42** catalyzed dearomative [3+2] cycloaddition of vinyl cyclopropane **27** to nitroindoles.

In a similar approach, Veselý and coworkers reported in 2019 a Pd-catalyzed [3+2] cycloaddition of vinyl cyclopropane azalactones with aldehydes activated by means of organocatalysis *(84)*. In this case, the selectivity of the transformation is determined by the selection of the organocatalyst since the stereoselectivity outcome of the reaction is determined on the ring-closing step and the Pd(0) only facilitates de opening of the vinyl cyclopropane moiety (Scheme 65). After screening of reaction conditions, compound **7** was selected as optimal organocatalyst yielding good diastereoselectivities and excellent enantioselectivities (up to 4:1 dr and up to 99% ee; Scheme 65).

In 2018, Rios and coworkers used the same strategy to obtain cyclopentanes through an unusual formal Pd-catalyzed [6+2] ring contraction of oxepino-fused pyrazolones with α,β-unsaturated aldehydes to obtain chiral cyclopentane-fused spiropyrazoles (Scheme 66) *(85)*. Again, the use of organocatalyst **7** provided the best diastereo- and enantioselectivities for a

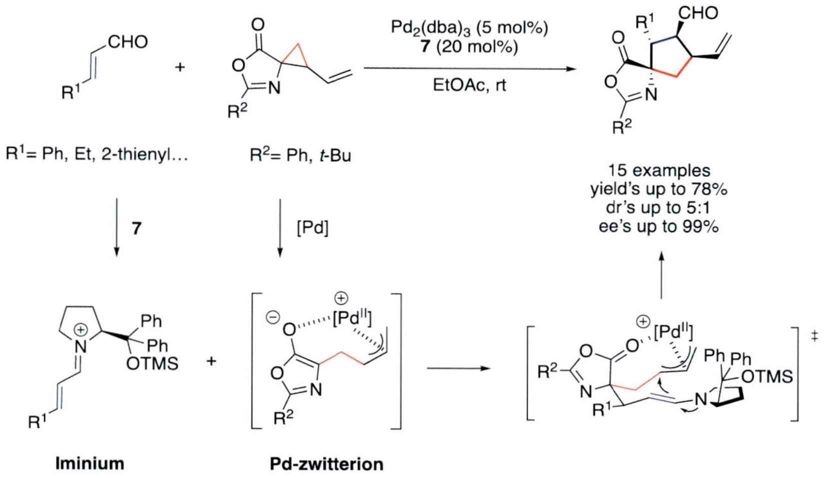

Metal-π-allyl mediated asymmetric cycloaddition reactions | 147

Scheme 65 Synthesis of indole-fused cyclopentanes via bifunctional organocatalysis and Pd-catalyzed [3 + 2] cycloaddition of vinyl cyclopropane azalactones with activated aldehydes and selectivity step intermediate.

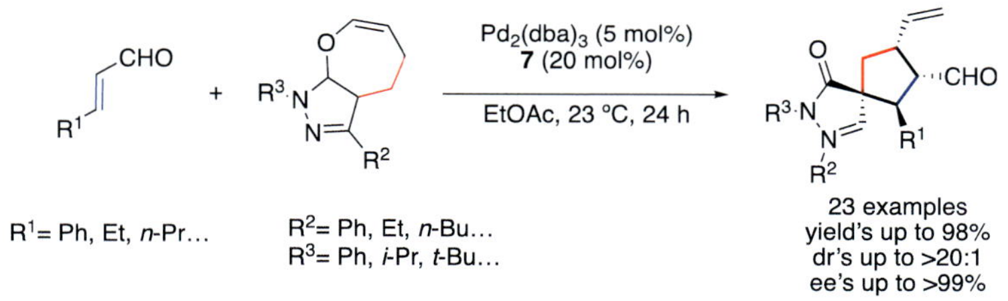

Scheme 66 Synthesis of cyclopentane-fused spiropyrazoles via organocatalysis and Pd-catalyzed [6 + 2] ring contraction/cycloaddition of oxepino-fused pyrazolones with α,β-unsaturated aldehydes.

variety of pyrazoles and α,β–unsaturated aldehydes (up to >20:1 dr and up to >99% ee).

Later, Liu and coworkers developed a synthetic method to obtain chiral cyclopentanes through the formal Pd–catalyzed [3 + 2] cycloaddition of vinyl cyclopropane **11** with cyclic oxathiazine dioxide–derived 1-azadienes *(86)*. Ligand screening disclosed phosphoroamidite *(R,R,R)*-**L27** as the optimal ligand for this transformation, reaching excellent diastereo– and enantio-selectivities independently of the steric and electronic properties of both aryl groups in the 1–azadiene substrate (up to 20:1 dr and up to 98% ee; Scheme 67). Scale up experiments and synthetic transformation were carried out to further demonstrate the utility of the catalytic methodology *(86)*.

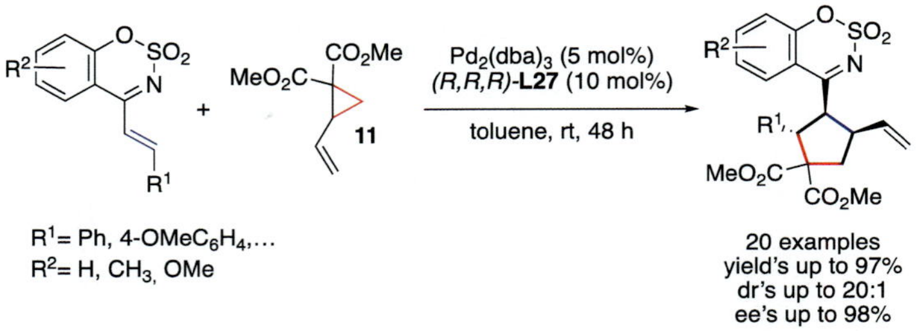

Scheme 67 Pd/(*R,R,R*)-**L27** catalyzed [3 + 2] cycloaddition of vinyl cyclopropane **11** with cyclic derived 1-azadienes.

Later in 2020, Deng and coworkers proposed a catalytic transformation for the synthesis of indole-fused cyclopentanes that involved the Pd-catalyzed [3 + 2] decarboxylative cycloaddition of vinyl indoloxazolidinones with electron deficient alkenes *(77)*. Optimization experiments carried out with malononitrile derivatives yielded excellent diastereo- and enantio-selectivities by using phosphoroamidite ligand (*R,R*)-**L43** (up to >20:1 dr and up to >99% ee; Scheme 68A). It is to note that regioselectivity of

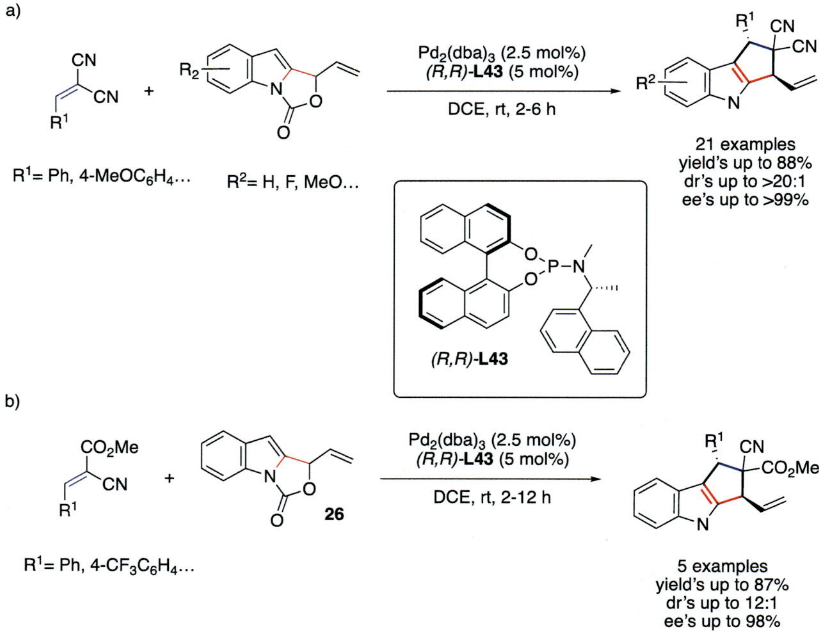

Scheme 68 Pd/(*R,R*)-**L43** catalyzed [3 + 2] decarboxylative cycloaddition of vinyl indoloxazolidinones with (A) malonitrile derivatives and (B) 2-cyano-3-phenylacrylates.

the reaction was perfectly controlled by the ligand obtaining mainly the cyclopentane[*b*]indole derivatives. In addition, similar excellent results were obtained using indoloxazolidinone **26** with a range of 2-cyano-3-phenyl acrylates (up to 12:1 dr and up to 98% ee; Scheme 68B) *(77)*.

The same year, Du and coworkers developed a synergistic Pd(0)/Rh(III) catalytic system for the formal [3+2] cycloaddition of spirovinyl cyclopropanes to α,β-unsaturated 2-acyl imidazoles *(87)*. In this respect the Pd(0) species are responsible of generating a stabilized zwitterionic π-allyl palladium (via the ring opening of the spirovinyl cyclopropanes) while the chiral Rh(III) complex activates the α,β-unsaturated 2-acyl imidazoles via coordination and controls the selectivity of the process (Scheme 69). Several Rh(III) catalyst were screened and it was found out that complex (*R*)-**C4** provided the optimal diastereo- and enantioselectivities (up to 14.5:1 dr and up to 99% ee; Scheme 69). Based on the stereochemical outcome of the reaction, the authors propose a plausible transition state in which the *t*-butyl groups of the (*R*)-**C4** complex block the *Si*-face of the intermediate and therefore prompts the *Re*-face attack (Scheme 69). Gram scale transformation and synthetic derivatizations were carried out on the newly prepared chiral spirocyclopentanes in order to demonstrate the utility of the protocol *(87)*.

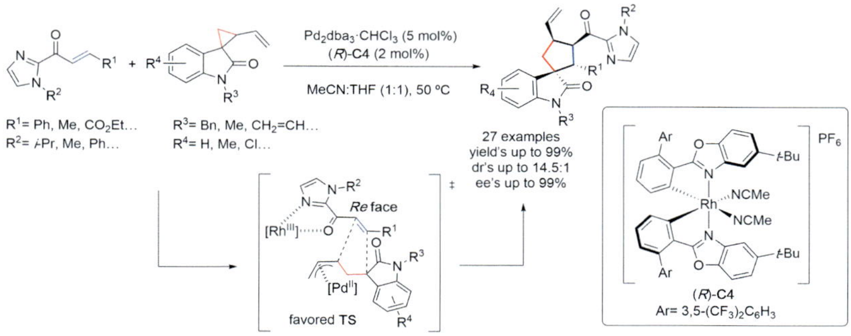

Scheme 69 Synergystic Pd/(*R*)-**C4** catalyzed [3+2] spirovinyl cyclopropanes and α,β-unsaturated 2-acyl imidazoles.

More recently, Li and coworkers synthetized chiral spirocyclopentanes scaffolds through a Pd-catalyzed [3+2] cycloaddition of activated vinyl cyclopropanes with benzofuran-derived azadienes *(88)*. Ligand screening revealed that pyrrolidine-derived phosphoroamidite ligand (*R,S*)-**L44** was necessary to obtain the best diastereo- and enantioselectivities (up to >20:1 dr and up to >99% ee; Scheme 70A). Interestingly, it was possible

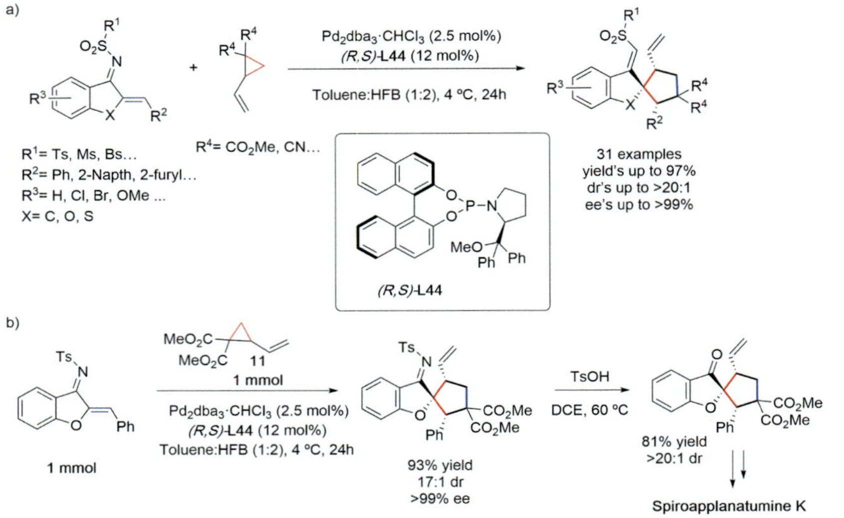

Scheme 70 (A) Pd/(R,S)-**L44** catalyzed [3 + 2] cycloaddition of activated vinyl cyclopropanes with benzofuran-derived azadienes. (B) Gram scale synthesis applied to the preparation of the core of spiroapplanatumine K.

to substitute the heteroatom of the benzofuranone scaffold (e.g., sulfur, nitrogen) as well as the electron withdrawing groups in the VCPs substrate without erosion of the selectivity *(88)*. Gram–scale synthesis and further transformations of the spirocyclopentane compounds was performed, providing a feasible method for the synthesis of the core of spiroapplanatumine K (Scheme 70B) *(89)*.

Another approach to obtain highly decorated cyclopentanes in an enantioselective way is through the Pd-catalyzed [3 + 2] cycloaddition reaction of trimethylenemethane (TMM) to a diverse range of electron deficient olefins *(90)*. In 2018, Trost and coworkers applied the Pd–TMM catalyzed cycloaddition of the cyano-methylallyl derivate **28** to β–nitroimine acceptor **29** for the total synthesis of the aminocyclitol core of jogyamycin (compound **30**) *(91)*. After ligand screening, the use of diamidophosphite ligand (S,S,R,R,S,S)-**L11** yielded the desired cycloadduct in high diastereo- and enantioselectivity (13:1 dr and 89% ee; Scheme 71).

Soon later, the same group developed another TMM-based Pd-catalyzed [3 + 2] cycloaddition reaction for the construction of chiral cyclopentane units via the reaction of imino–methyallyl donor **17** with β–nitroolefins (Scheme 72A) *(92)*. Best diastereo- and enantioselectivities were obtained using phosphoroamidite ligand (S,S,S)-**L9** (up to >20:1 dr and up to

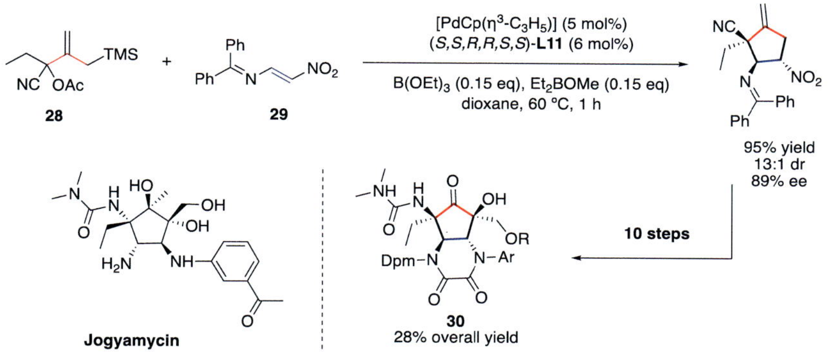

Scheme 71 Synthesis of the aminocyclitol core of jogyamycin via Pd-TMM catalyzed [3 + 2] cycloaddition reaction.

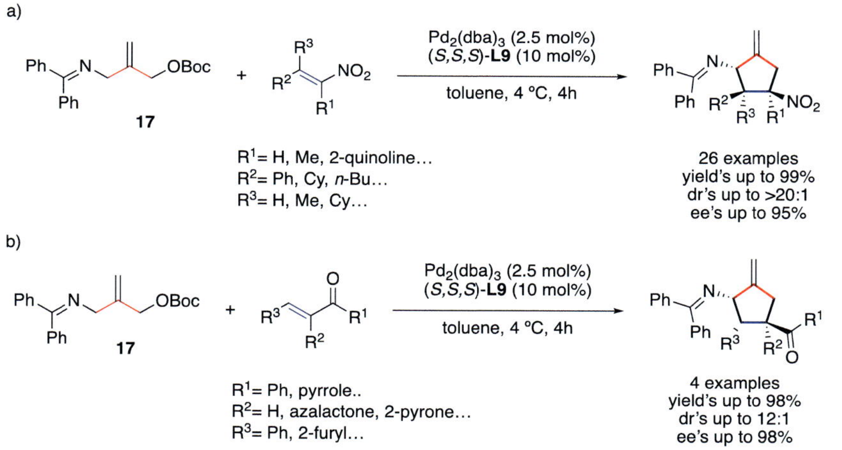

Scheme 72 Pd/(*S*,*S*,*S*)-**L9** catalyzed [3 + 2] cycloaddition of imino-methylallyl derivative **17** with (A) β-nitrololefins and (B) to α,β-unsaturated carbonylic compounds.

95% ee; Scheme 72A). The use of other electrodeficient olefins (e.g., α,β–unsaturated ketones, azalactones, 2–pyrones, etc.) rather than β–nitroolefins, provided the desired cycloadducts under the same conditions in slightly lower diastereoselectivities but still high enantioselectivities (up to 12:1 dr and up to 98% ee; Scheme 72B). One–pot functionalization and post–functionalitzacion were carried out to prove the synthetic utility of the protocol, accessing a cycloadduct that can be transformed into the patented flu drug by Abbot *(92,93)*.

The same year, Trost and coworkers developed the Pd-catalyzed [3 + 5] cycloaddition of allenyl-methylallyl donors with nitroolefin-imino substrate **29** *(94)*. Ligand optimization revealed that the use of diamidophosphite ligand (*R,R,S,S,R,R*)-**L45** provided the best diastereo- and enantio-selectivities in a range of different allenyl-methylallyl donors (up to >20:1 dr and up to 99% ee; Scheme 73A). Mechanistic studies revealed that the presence of a dynamic kinetic asymmetric transformation (DKYAT) that took place in the Pd–TMM intermediate *(94)*. It is proposed that the steric hindrance of the Pd complex and the R^1-group in the allene substrate prompts the π–σ–π equilibration that places both substituents in opposite faces

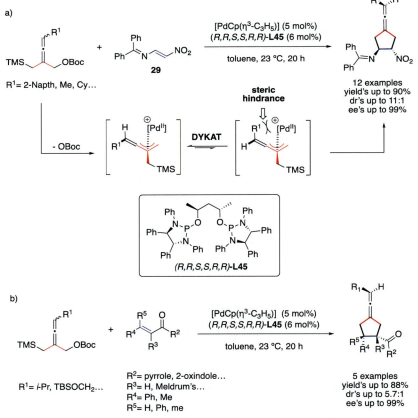

Scheme 73 Pd/(*R,R,S,S,R,R*)-**L45** catalyzed [3 + 2] cycloaddition allenyl-methylallyl derivatives with (A) compound **29** and (B) α,β-unsaturated carbonylic compounds.

(Scheme 73A). The same catalytic conditions could be applied with other acceptor substrates (e.g., N-acyl pyrroles, oxindoles, Meldrum-derivatives, etc.) without significant erosion of the selectivity (up to 5.7:1 dr and up to 98% ee; Scheme 73B). The synthetic utility of the protocol was demonstrated by applying the allenyl-cycloadducts to a variety of different transformations like: regioselective allene-alkene coupling, Rh-catalyzed C-H activation, synthesis of spirocycles and allylic alkylation *(94)*.

In 2019, Trost and coworkers extended the Pd-TMM methodology to the synthesis of complex chiral heteroaryl cyclopentanes. In this case the formal Pd-catalyzed [3+2] cycloaddition was performed with several heterocyclic-containing methylallyl donors and α,β-unsaturated carbonylic compounds *(95)*. The use of diamidophosphite ligand (S,S,R,R,S,S)-**L11** yielded the heteroaryl-decorated cycloadducts (e.g., pyridine, quinoline, pyrimidine, etc.) in high diastereo- and enantioselectivities (up to >20:1 dr and up to 99% ee; Scheme 74A). In addition, similar high diastereo- and enantioselectivity were attained (up to >20:1 dr and up to 98% ee; Scheme 74B) when using other cyclic and acyclic acceptor substrates (e.g., nitroalkenes, sulfones, azalactones, etc.). Again, the utility of the synthetic protocol was demonstrated applying the cycloadducts in several post-functionalization reactions such as cyclopropanation of the double bond, regioselective oxidation or diastereoselective dihydroxylation *(95)*.

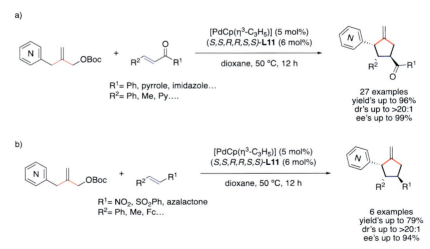

Scheme 74 Pd/(S,S,R,R,S,S)-**L11** catalyzed [3+2] cycloaddition heteroaryl-methylallyl derivatives with (A) α,β-unsaturated carbonylic compounds and (B) other acceptor olefins.

In 2020, the same group developed a Pd–mediated synthesis of chiral cyclopentyl organophosphonates through the formal [3 + 2] cycloaddition of phosphonate–methylallyl donors to β–nitroolefins *(47)*. The use of Pd/ (S,S,R,R,S,S)–**L11** catalytic system provided the best results (up to >20:1 dr and up to 99% ee; Scheme 75). They also demonstrated that other activated alkenes rather than β–nitroolefins (e.g., sulphones, barbiture derivatives, etc.) can also be efficiently used as starting materials in this transformation (up to >20:1 dr and up to 99% ee) *(47)*.

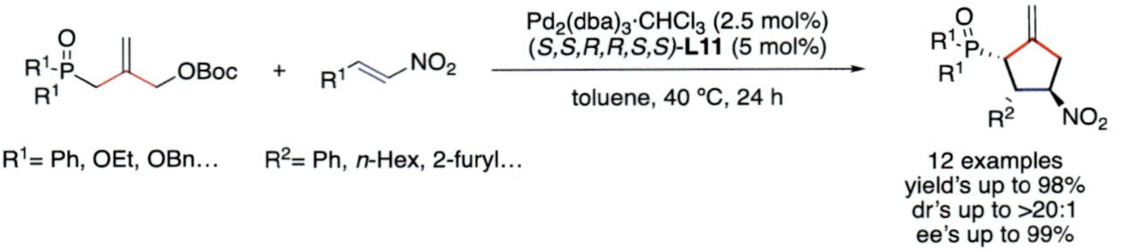

Scheme 75 Pd/(S,S,R,R,S,S)–**L11** catalyzed [3 + 2] cycloaddition phosphonate-methylallyl derivatives with β-nitroolefins.

Later the same year, Trost and coworkers used the Pd–TMM approach for synthesis of chiral trifluoromethyl–containing cycloadducts via Pd–catalyzed [3 + 2] cycloaddition of CF_3–decorated methylallyl donor **31** with electrodeficient olefins *(21)*. The use of ligand (S,S,R,R,S,S)–**L11** yielded the best selectivity results (up to >20:1 dr and up to 99% ee; Scheme 76A).

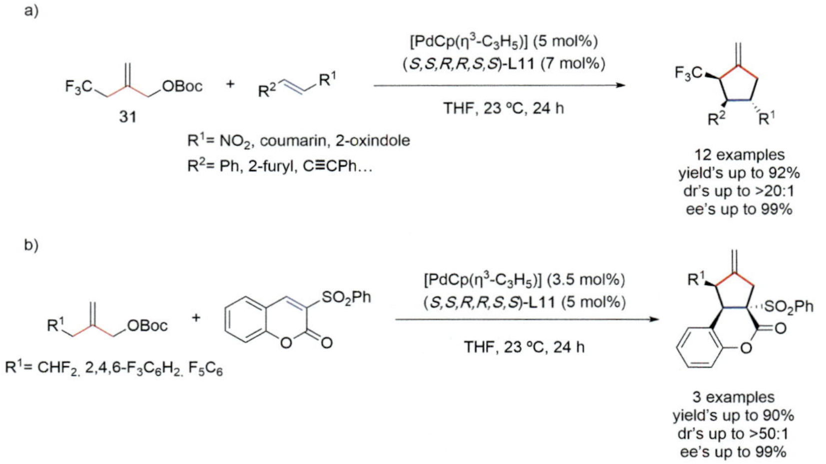

Scheme 76 Pd/(S,S,R,R,S,S)–**L11** catalyzed [3 + 2] cycloaddition of (A) CF_3-methylallyl donor **31** with various acceptor substrates and (B) fluorine-containing methylallyl substrates with phenylsulfonyl-functionalized coumarin.

Substitution of the trifluoromethyl moiety for other fluorinated groups (e.g., difluoromethyl) as well of fluorinated aryl moieties (e.g., 2,4,6-trifluorophenyl and pentafluorophenyl) were also feasible reaching comparable selectivities to the trifluoromethylated analogs (up to >50:1 dr and up to 99% ee; Scheme 76B).

The same group discovered in 2020 an unusual reactivity pattern in the Pd-catalyzed [3 + 2] cycloaddition of methylallyl donors in which the *in situ* generated 1,4-dipoles undergo a nucleophilic attack in the C2 of the Pd-allyl (Scheme 77) *(96)*. Ligand selection was essential to maximize the regioselective ring closing and synthesis of the spirofused cyclopentanes in high diastereo- and enantioselectivities. In this case the best results were obtained with Feringa-type phosphoroamidite (*S,S,S*)-**L46** (up to >15:1 dr and up to 91% ee; Scheme 77). This kind of spirofused cycloadducts are present in many several bioactive molecules such as Flaviviridae virus inhibitors *(97,98)*.

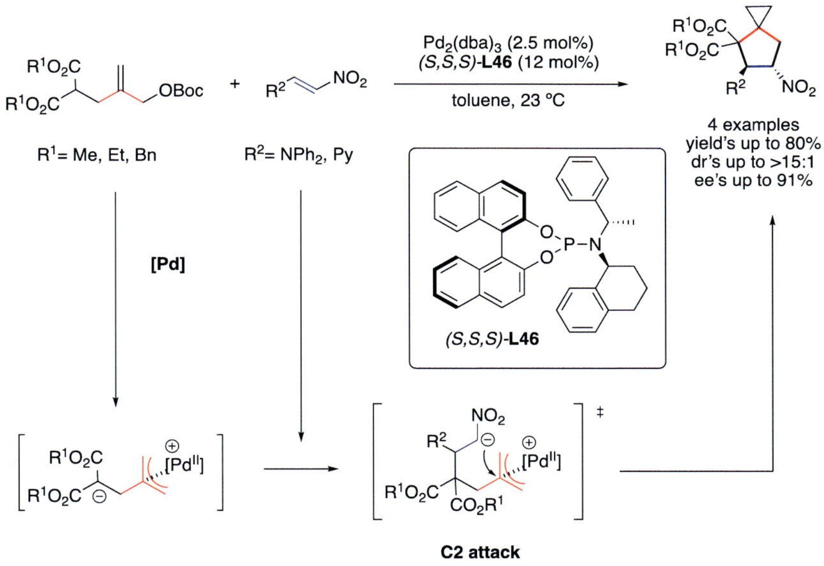

Scheme 77 Pd/(*S,S,S*)-**L46** catalyzed [3 + 2] cycloaddition of methylallyl donors with β-nitroolefins.

In 2021, Trost and coworkers disclosed another Pd-catalyzed [3 + 2] cycloaddition reaction involving 2-acyl-methylallyl donor substrates and electrodeficient olefins to easily access chiral 2-acyl-cyclopentanes *(99)*. Optimization lead to the conclusion that the use of diamidophosphite

(R,R,S,S,R,R)-**L45** provided the best results in a variety of 2-acyl methylallyl donors as well for the acceptor nitroalkenes reaching high diastereo- and enantioselectivities (up to >19:1 dr and up to 99% ee; Scheme 78). Remarkably this synthetic approach was also succesful with other acceptor molecules such as: azalactones, malonitriles and barbiture-derived olefins without erosion of selectivity (up to >19:1 dr and up to 98% ee) *(99)*.

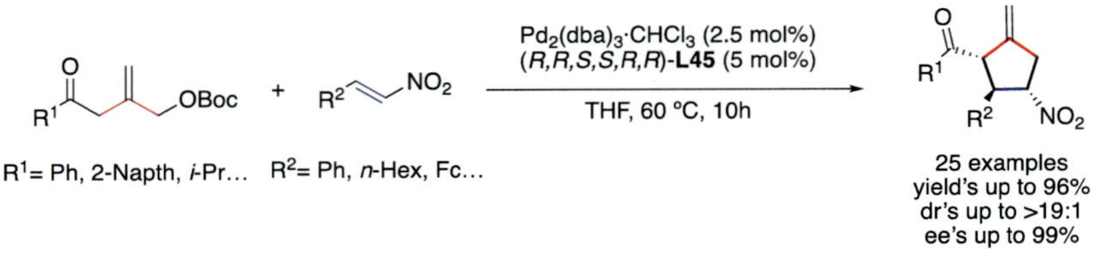

Scheme 78 Pd/(R,R,S,S,R,R)-**L45** catalyzed [3 + 2] cycloaddition 2-acyl-methylallyl donor substrates to nitroolefins.

Another interesting strategy to obtain chiral cycloadducts is using a zwit-terionic Pd–oxyallyl species *(100)*. Zi and coworkers reported in 2021 the use of this species generated from vinyl methylene cyclic carbonates in an unprecedented Pd-catalyzed [3 + 2] cycloaddition with nitroalkenes to form chiral cyclopentanones *(101)*. Development of the new hydrogen-bond-donating ligand (R,S)-**L47** was necessary maximize the diastereo- and enantioselectivities (up to >20:1 dr and up to >99% ee; Scheme 79). The role of the newly synthetized (R,S)-**L47** ligand is vital to the enantioselectivity outcome of the reaction since the urea tether forms hydrogen bonds with both the oxygen of the Pd–oxyallyl complex as well as with the nitro group that directs the nucleophilic attack (Scheme 79).

4.2 Cyclopentenes

Highly decorated chiral cyclopentenes and their derivatives are recurrent structural subunits in biologically active natural products *(102)* and drug-like substances *(103)*. Hou and coworkers developed a simple protocol for the synthesis of chiral cyclopentenes through a formal Pd-catalyzed [3 + 2] cycloaddition of VCP **3** with electron deficient alkynes *(104)*. Optimization of the reaction conditions proved that the optimal ligand was (R)-SegPhos yielding selectively the cyclopentane adducts in good enantioselectivities for a range of different alkynyl α-ketoesters (up to 89% ee; Scheme 80). The catalytic protocol could also be applied to alkynyl 1,2-diones with a similar selectivity outcome (up to 85% ee; Scheme 80).

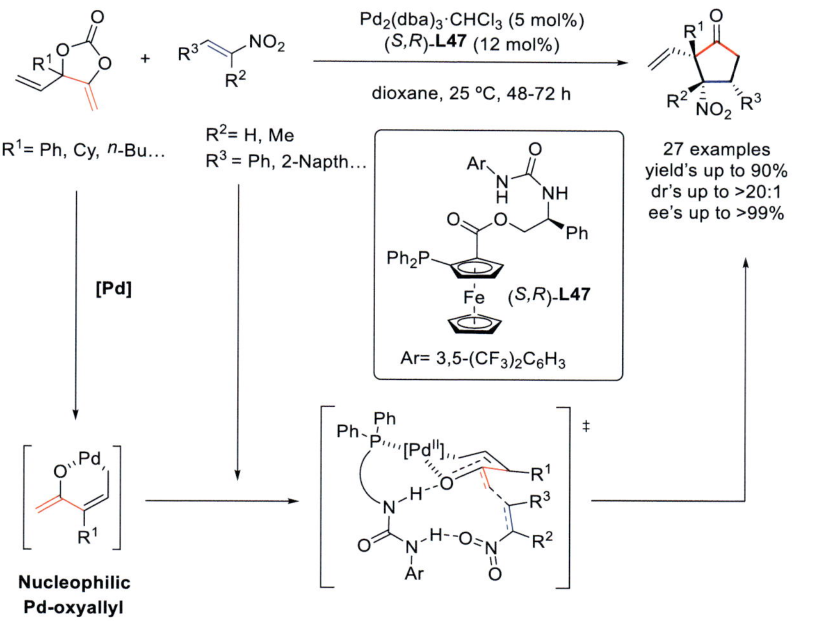

Scheme 79 Pd/(R,S)-**L47** catalyzed [3+2] cycloaddition of vinyl methylene cyclic carbonates to β-nitroolefins.

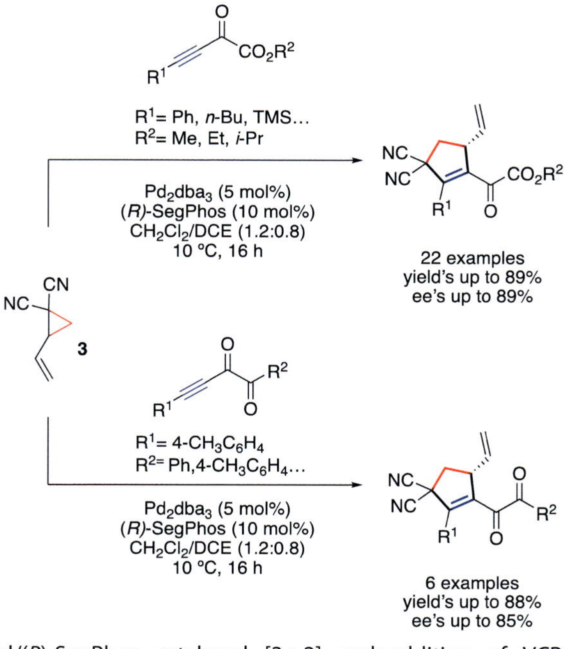

Scheme 80 Pd/(R)-SegPhos catalyzed [3+2] cycloaddition of VCP **3** to alkynyl α-ketoesters and alkynyl 1,2-diones.

4.3 Cyclohexanes

Six-membered chiral cycloadducts are present in many natural recurring structures as well as bioactive molecules *(105)*. One of the most effective ways to prepare this compound is trough Pd-catalyzed [4 + 2] cycloaddition reactions involving the ring opening of cyclic compounds *(106)*. For instance, Fan and coworkers reported in 2021 the Pd-catalyzed [4 + 2] cycloaddition of γ-methylidene-δ-valerolactones (GMDVs) with isatin-derived *p*-quinone methides to yield highly decorated spirooxindoles-fused cyclohexanes *(107)*. The use of spirobisphosphine ligand (*S,S,S*)-**L48** provided the best diastereo- and enantioselectivities for a range of different *p*-quinone methides and GMDVs (up to >20:1 dr and up to 95% ee; Scheme 81). Diastereoselectivity outcome of the reaction led to the hypothesis that the transition state of the selectivity determining step must involve a *Si*-face to *Si*-face attack due to the C_2 symmetry of ligand (*S,S,S*)-**L48** (Scheme 81). Gram scale synthesis and post functionalization was performed to demonstrate the synthetic utility of the protocol *(107)*.

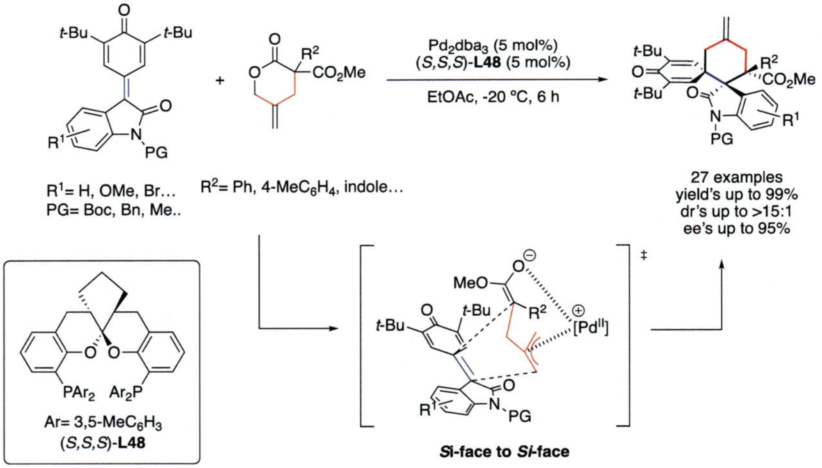

Scheme 81 Synthesis spirooxindoles-fused cyclohexanes via Pd/(*S,S,S*)-**L48** catalyzed [4 + 2] cycloaddition of γ-methylidene-δ-valerolactones with isatin-derived *p*-quinone methides.

Another approach to synthetize chiral 6-membered cycloadducts is through the Pd-catalyzed [4 + 2] cycloaddition reaction of trimethylenemethane (TMM) to a diverse set of electron deficient olefins *(108)*. For example, in 2020, Trost and coworkers developed a Pd-catalyzed [4 + 2] cycloaddition reaction involving methylallyl donor **29** with *in situ* generated 1,4-dipoles *(96)*. Ligand selection was crucial to obtain exclusively the six-membered ring. In this respect, phosphoroamidite ligand (*S,S,S*)-**L49** gave simultaneously the best regio-, diastereo- and enantioselectivity (up to 20:1 dr and up to 95% ee; Scheme 82A). When other acceptor molecules were tested a change of ligand was necessary to obtain the best selectivities. Thus, on one hand, the use of ligand (*S,R,R*)-**L9** was required for methylallyl acceptors containing an heterocyclic substituent (e.g., pyridine, quinoline, benzimidazole, etc.) as electron withdrawing group (up to >15:1 dr and up to 95% ee; Scheme 82A). On the other hand, the use of diamidophosphite (*S,S,R,R,S,S*)-**L11** was necessary for 2-oxindole derived TMMs (up to 5.3:1 dr and up to 90% ee; Scheme 82B), whereas for the cycloaddition of *N*-isatins derived TMMs to different acceptor molecules the use of ligand (*R,R,S,S,R,R*)-**L45** was required (up to >15:1 dr and up to 95% ee; Scheme 82B) *(96)*.

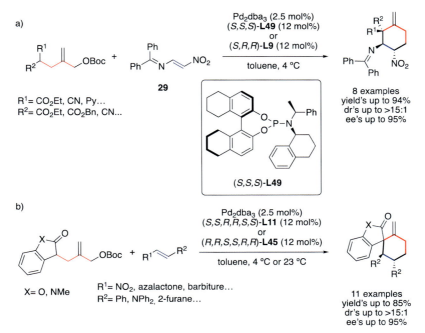

Scheme 82 (A) Pd-catalyzed [4+2] cycloaddition of malonate- and heterocyclic-methylallyl donors with imino-nitroalkene **29**. (B) Pd- catalyzed [4+2] cycloaddition of 2-oxindole- and *N*-isatin derived methylallyl donors to different Michael acceptors.

In the same year, Trost and coworkers also reported a Pd-catalyzed [4+2] cycloaddition reaction involving 1,4-dipoles, generated in situ from α-carbonyl-methylallyl substrates, and electrodeficient olefins to form highly decorated chiral cyclohexanones *(109)*. Ligand optimization revealed that the use of pyrrolidone-containing phosphoroamidite ligand (*R,S,S*)-**L50** provided the best diastereo- and enantioselectivities (up to >15:1 dr and up to 95% ee; Scheme 83A). Gram-scale synthesis of the cycloadducts, which can be used to synthetize precursors of bioactive molecules such as Ganoderma aldehyde *(110)* (Scheme 83B) or Canangone *(111)*, further demonstrated the convenience of the protocol.

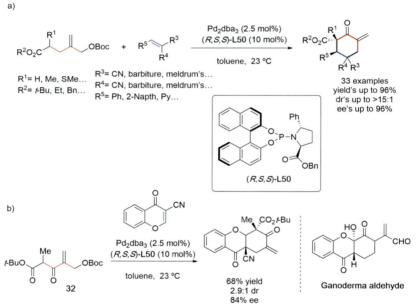

Scheme 83 (A) Pd/(*R,S,S*)-**L50** catalyzed [4+2] cycloaddition of carbonyl-methylallyl donnors and electron deficient alkenes. (B) Synthesis of Ganoderma aldehyde related compound via Pd/(*R,S,S*)-**L50** catalyzed [4+2] cycloaddition.

4.4 Cyclohexenes

Spiro-fused unsaturated six-membered rings are present in a diverse range of bioactive natural and synthetic molecules *(112)*. Zhao and coworkers reported in 2018 a Pd/Ti relay catalyzed [4+2] cycloaddition reaction of vinylethylene carbonates with aurones *(113)*. In order to afford these spirofused cyclohexenes the addition of a Ti(O*i*-Pr) switches the Pd-Zwitterion from the alkoxide-π-allyl to a dienolate. Reaction of the later with (*Z*)-2-benzylidenebenzofuran-3(2*H*)-one yielded the corresponding cyclohexene compound (Scheme 84). High yields and high diastereo-selectivities were achieved, however, only one asymmetric example was reported. The cycloaddition of 1,3-indandione **33** with 4-phenyl-4-vinyl-1,3-dioxolan-2-one using achiral 1,2-bis(diphenylphosphino)benzene (dppbz) and (−)-TADDOL gave the cycloadduct in high diastereoselectivity (>20:1 dr) and only 60% ee (Scheme 84). These results provided support to the theory the Ti-enolate species are involved in the stereodetermining step of the reaction instead of the Pd-zwitterionic complex *(113)*.

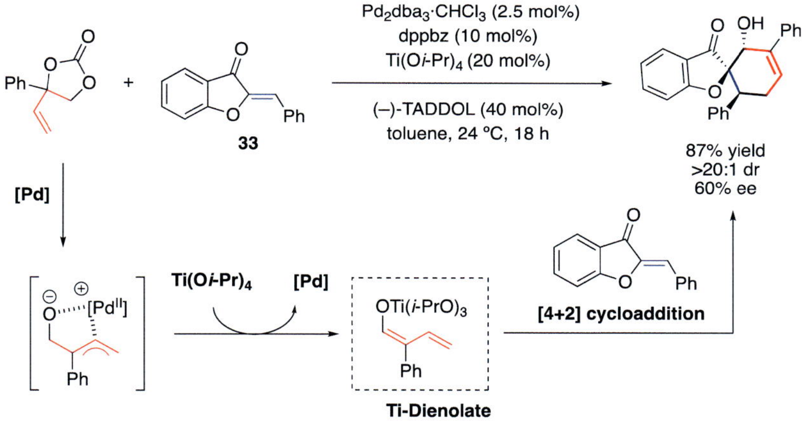

Scheme 84 Synthesis of spirofused cyclohexene via Pd/Ti relay catalyzed [4 + 2] cyclo-addition of 1,3-indandione **33** with 4-phenyl-4-vinyl-1,3-dioxolan-2-one.

4.5 Cycloheptanes

Cyclohepta[*b*]indoles take part in several motives of bioactive natural products *(114)*. However the construction of seven membered rings via Pd-catalyzed [4 + 3] cycloaddition is underdeveloped. Deng and coworkers pioneered in this field by reporting in 2020 one of the first all carbon Pd-catalyzed [4 + 3] cyclo-additions of TMM precursor **4** to indole-2,3-quinomethanes *(115)*. This synthetic method afforded various cyclohepta[*b*]indoles in high diastereo- and enantioselectivities (up to >20:1 dr and up to >99% ee; Scheme 85A) when using tetrahydroquinoline-based phosphoroamidite ligand (*S*)-**L51**. The same methodology was extended to the use of pyrrolidone-3,4-diene substrates with

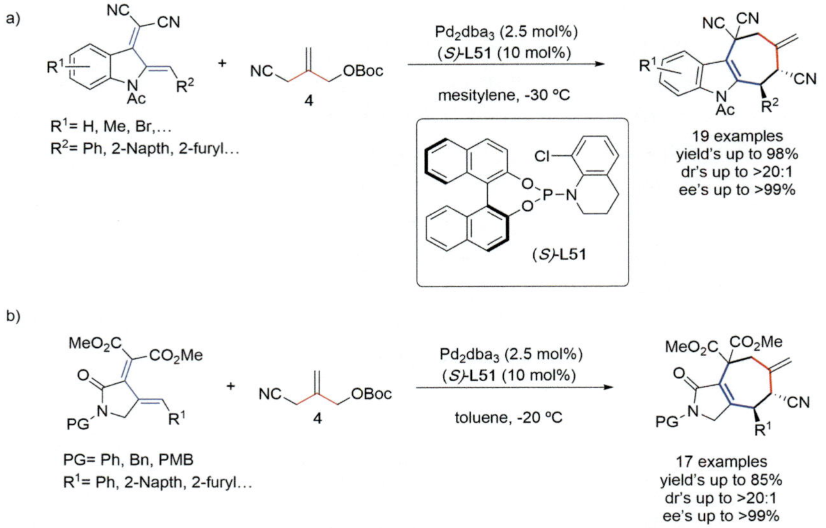

a)

R^1= H, Me, Br,...
R^2= Ph, 2-Napth, 2-furyl...

Pd$_2$dba$_3$ (2.5 mol%)
(S)-**L51** (10 mol%)

mesitylene, -30 °C

(S)-**L51**

19 examples
yield's up to 98%
dr's up to >20:1
ee's up to >99%

b)

PG= Ph, Bn, PMB
R^1= Ph, 2-Napth, 2-furyl...

Pd$_2$dba$_3$ (2.5 mol%)
(S)-**L51** (10 mol%)

toluene, -20 °C

17 examples
yield's up to 85%
dr's up to >20:1
ee's up to >99%

Scheme 85 (A) Synthesis of cyclohepta[*b*]indoles via Pd/(S)-**L51** catalyzed [4+3] cyclo-addition of TMM precursor **4** with indole-2,3-quinomethanes. (b) Synthesis of cyclohepta[*c*]pyrrol-1(2*H*)-ones via Pd/(S)-**L51** catalyzed [4+3] cycloaddition of TMM precursor **4** with pyrrolidone-3,4-dienes.

a malonate group instead of a malonitrile. This simple modification afforded the corresponding cyclohepta[*c*]pyrrol-1(2*H*)-ones in high diastereo- and enantioselectivities (up to >20:1 dr and up to >99% ee; Scheme 85B).

4.6 Bicyclodecadienes

Ketone-containing polycyclic structures are recurrent in several bioactive and natural products motives *(116)*. In this context, Trost and coworkers found out that tropone derivatives can undergo Pd-catalyzed [6+3] cyclo-addition reactions with different methylallyl donors. For instance, imino-methylallyl donor **29** underwent cycloaddition with tropone derivative **34** in the presence of ligand (*R,S,S*)-**L9** yielding high diastereo- and high enantioselectivity (8:1 dr and 95% ee; Scheme 86A) *(92)*. The reaction of tropone **35** with allene methylallyl donor **36** yielded a mixture of *meso* iomers (2.3:1 dr, Scheme 86B) by using ligand (*R,R,S,S,R,R*)-**L45** *(94)*. Interestingly, the cycloaddition of tropone **35** with imidazole methylallyl donor **37** afforded the corresponding bicyclo[4.3.1]decadiene in good diastereo- and enantioselectivity (4:1 dr and 85% ee; Scheme 86C) using ligand (*S,S,R,R,S,S*)-**L11** *(95)*.

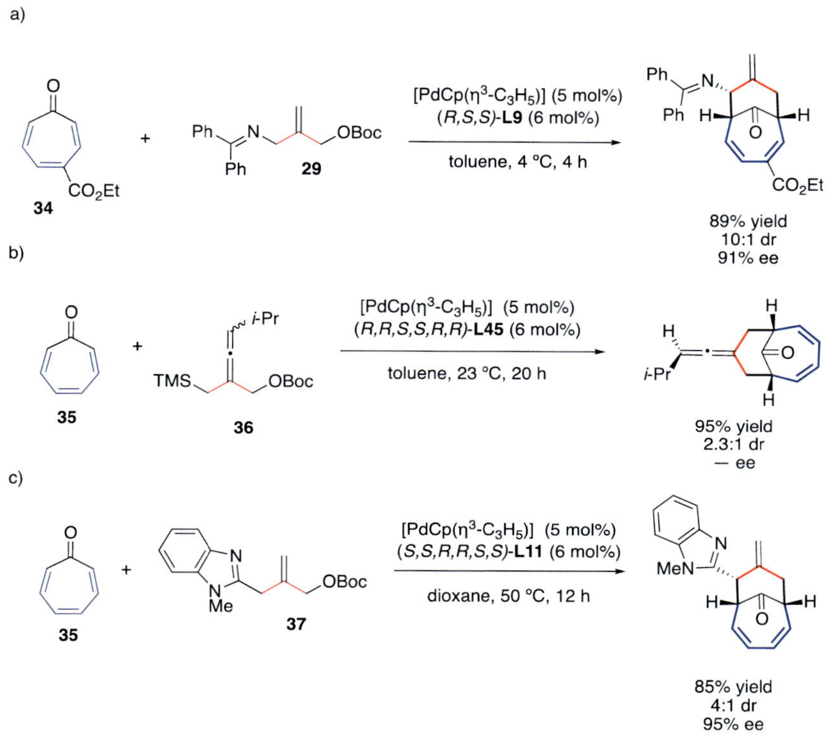

Scheme 86 (A) Pd/(R,S,S)-**L9** [6+3] catalyzed cycloaddition of TMM precursor **29** to tropone-derivate **34**. (B) Pd/(R,R,S,S,R,R)-**L45** catalyzed [6+3] cycloaddition of TMM precursor **36** to tropone-derivate **35**. (C) Pd/(S,S,R,R,S,S)-**L11** catalyzed [6+3] cycloaddition of TMM precursor **37** to tropone-derivate **35**.

5. Cycloaddition reactions for the formation of mixed-heterocycles
5.1 Oxazoline-related compounds

Oxazoline-related compounds are present in several organic building blocks and natural products. Moreover, they are commonly used as chiral auxiliaries as well as ligands in metal–mediated catalysis *(117)*. In 2019, Zhang and coworkers reported on a method for the synthesis of isoxazolines *N*-oxides via the intramolecular Pd-catalyzed [3 + 2] cycloaddition reaction of nitro-containing allylic carbonates *(118)*. The latter were easily prepared via the Pd(PPh$_3$)$_4$ catalyzed allylic substitution of vinyl ethylene carbonates with arylated nitromethanes (Scheme 87). The authors observed that the

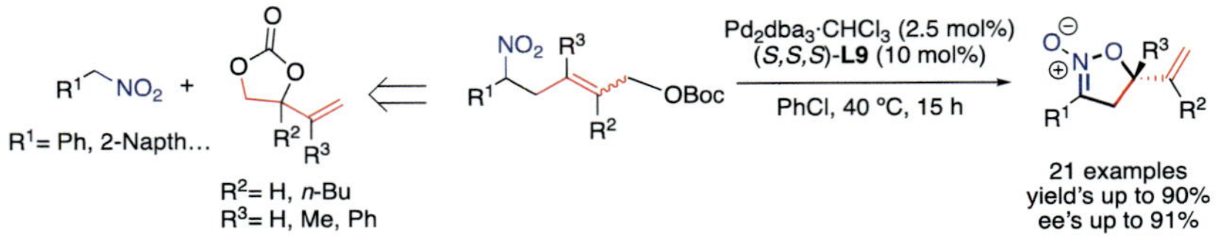

Scheme 87 Synthesis of isoxazolines N-oxides via the Pd-catalyzed intramolecular [3+2] cycloaddition reaction of nitro-containing allylic carbonates.

use of ligand (S,S,S)-**L9** provided high yields and good enantioselectivities (up to 91% ee; Scheme 87). Interesting compounds such as chiral amino alcohols and 2-hydroxy-pyrrolidines could be prepared from the newly formed isoxazolines N-oxides in gram-scale, which further demonstrated the usefulness of the protocol (118).

The same year, Ni and coworkers developed an asymmetric Pd-catalyzed [3+2] cycloaddition reaction of 2-vinyloxirane with N-benzothiazoloimines to yield chiral 2-oxazolidine compounds (46). Upon optimization, Pd/ (S)-BINAP catalytic system provided the 2-oxazolidine derivatives in high diastereo- and enantioselectivities (up to 8.5:1 dr and up to >99% ee; Scheme 88).

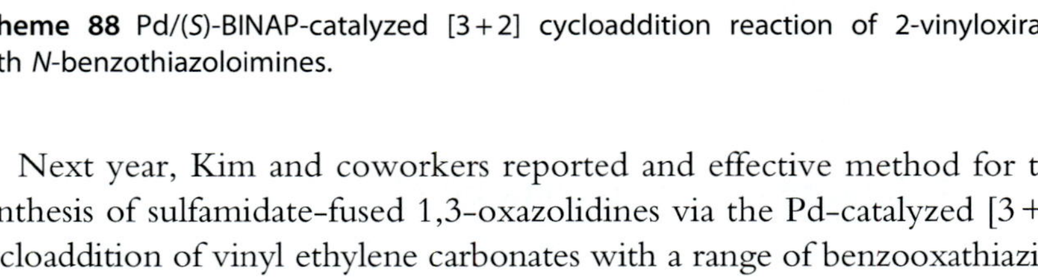

Scheme 88 Pd/(S)-BINAP-catalyzed [3+2] cycloaddition reaction of 2-vinyloxirane with N-benzothiazoloimines.

Next year, Kim and coworkers reported and effective method for the synthesis of sulfamidate-fused 1,3-oxazolidines via the Pd-catalyzed [3+2] cycloaddition of vinyl ethylene carbonates with a range of benzooxathiazine 2,2-dioxides (119). A variety of sulfamidate-fused 1,3-oxazolidines were prepared using phosphoroamidite (S,S,S)-**L9** in high diastereo- and enantio-selectivities (up to 25:1 dr and up to 99% ee; Scheme 89). Furthermore, the vinyl ethylene carbonates could be substituted by vinyl epoxides; however, diastereoselectivity and enantioselectivity notably decreased (119). Gram scale synthesis and post functionalization yielded interesting molecules such as chiral oxazolines which are central to many chiral ligand families (119,120).

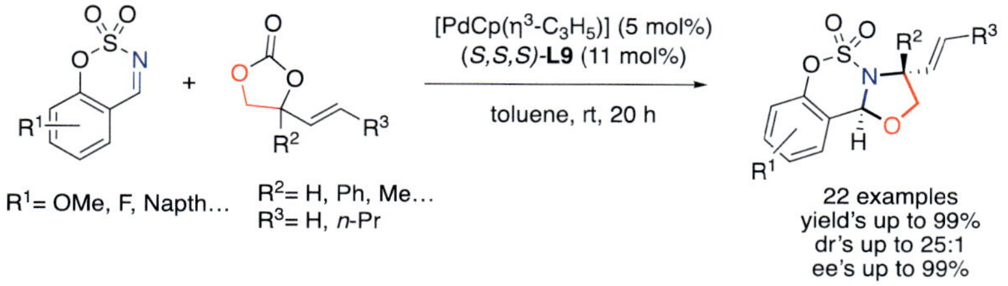

Scheme 89 Pd-catalyzed [3+2] cycloaddition of vinyl ethylene carbonates with benzooxathiazine 2,2-dioxides.

5.2 Oxazepines

Mixed heteroatom seven-membered rings are a very valuable pharmacophores and the oxazepine structure is present in many bioactive compounds *(121)*. In 2018, Shi and coworkers developed a synthetic method based in the [4+3] cycloaddition of *p*-quinone derivatives with vinyl aziridines to form benzo-fused oxazepines *(122)*. For this reaction, both Pd and Ir precursors were studied. Although both metal precursors yielded the desired seven-membered cycloadduct in good yields and diastereoselectivities, high ee's were only achieved using Pd/(*R,R*)-**L52** catalytic system (up to 10:1 dr and up to 92% ee, Scheme 90). Additionally, olefin metathesis could be performed in the newly synthetized cycloadducts whithout erosion in selectivity *(122)*.

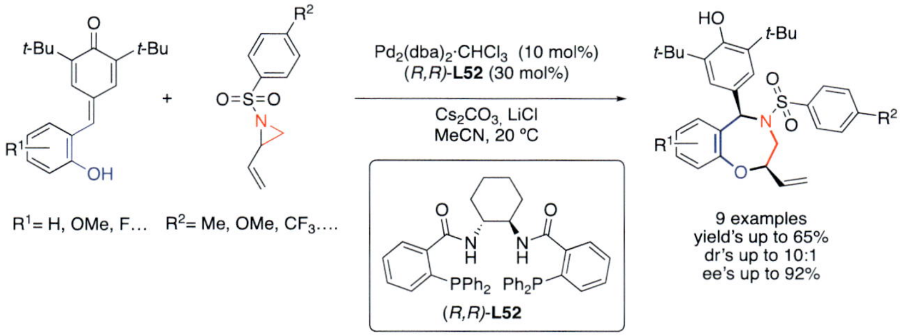

Scheme 90 Pd/(*R,R*)-**L52** catalyzed [4+3] cycloaddition of *p*-quinone derivatives with vinyl aziridines.

Later, Kim and coworkers disclosed the Pd-catalyzed [5+2] cycloaddition of *N*-benzothiazoloimines and vinyl ethylene carbonates in order to synthetize *N*-fused 1,3-oxepines *(123)*. The key in this transformation is the use of the ligand (*S*)-SegPhos, which is able to switch the previously reported [3+2] reactivity *(119)* to the [5+2] cycloaddition. High yields and good enantioselectivities were therefore achieved (up to 80% ee; Scheme 91).

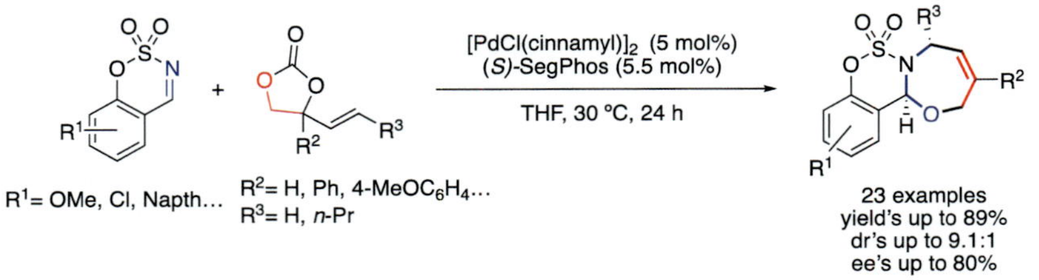

Scheme 91 Pd- catalyzed [5 + 2] cycloaddition of *N*-benzothiazoloimines and vinyl ethylene carbonates.

5.3 Epoxybenzoazepines

Several bioactive and pharmaceutical molecules contain the anthranil and its medium-sized ring derivatives *(124)*. You and coworkers developed a methodology to obtain cyclic epoxy-fused benzoxazepines via Pd-catalyzed [4 + 3] cycloaddition of anthranil derivatives with vinyl cyclopropanes *(125)*. The use of Pd/(*S*)-**L53** catalytic system was found to be optimal providing high diastereo- and enantioselectivities (up to 10:1 dr and up to 98% ee; Scheme 92). Based on control experiments, the addition of both Cs_2CO_3 and BEt_3 additives were found to be crucial for the reaction to occur. Additionally, ^1H and ^{11}B NMR mechanistic studies revealed the presence of a BEt_3-anthranil coordination adduct. Such adduct has been proposed to be responsible for the enhanced reactivity of the anthranil substrate (Scheme 92).

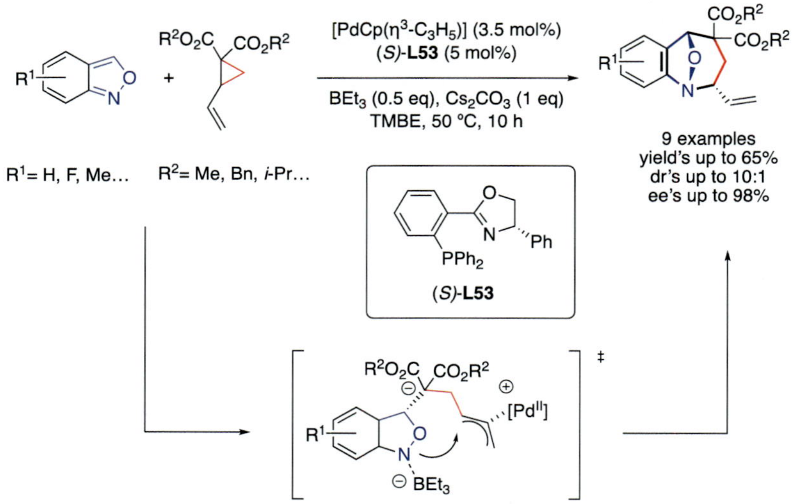

Scheme 92 Pd-catalyzed [4 + 3] cycloaddition of anthranil derivatives with vinyl cyclopropanes.

5.4 Oxazocines

Spirooxindole derived medium-sized heterocyclic cycloadducts have been found to be part of many bioactive and pharmaceutical molecules *(105d,126)*. In 2019, Shi and coworkers developed a formal Pd-catalyzed [5 + 3] cycloaddition between isatin-derived α-CF$_3$-imines and vinyl ethylene carbonates to form spirofused CF$_3$-containing–oxazocines *(127)*. Although several cycloadducts were obtained in high yields, only one enantioselective example was reported using ligand (*S,S,S*)-**L54** and Brønsted acid **39** (up to 63% ee; Scheme 93).

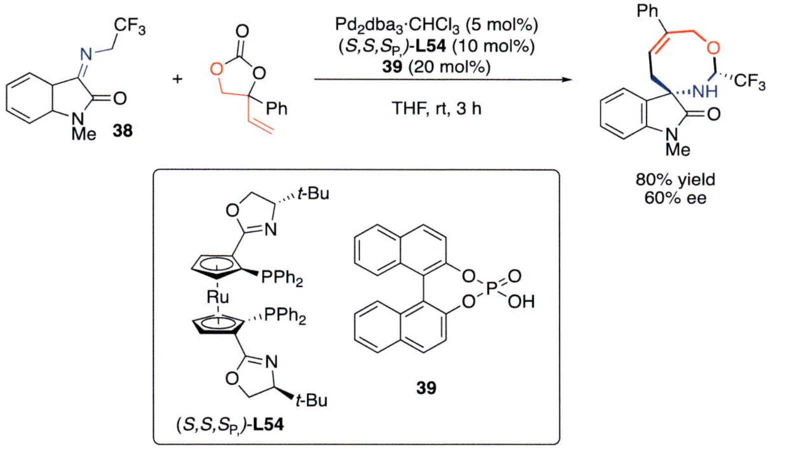

Scheme 93 Pd-catalyzed [5 + 3] cycloaddition between isatin-derived α-CF$_3$-imine **38** and 4-phenyl-4-vinyl-1,3-dioxolan-2-one.

5.5 Tetrahydrobenzoepoxyazocines

Eight-membered ring azocine derivatives are part of several pharmaceutical and bioactive products *(128)*. Li and coworkers recently reported a methodology based in the Pd-TMM catalyzed [4 + 4] cycloaddition between anthranils and γ-methylidene-δ-valerolactones for the synthesis of tetrahydrobenzoepoxyazocine cycloadducts *(129)*. The use of phosphoroamidite ligand (*S*)-**L55** provided the best diastereo- and enantioselectivities (up to 20:1 dr and up to 99% ee; Scheme 94A). Gram scale synthesis could be performed as well as the conversion of the newly formed cycloadducts into structural analogs as the antiarrhythmic agent analog shown in Scheme 94B *(129)*.

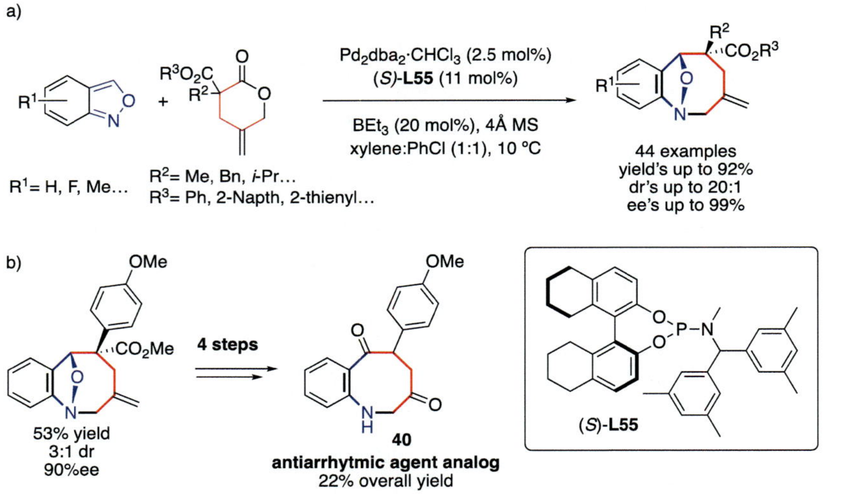

Scheme 94 (A) Pd-catalyzed [4+4] cycloaddition of anthranils with γ-methylidene-δ-valerolactones. (B) Synthesis of antiarrhythmic agent analog **40** from the newly synthetized cycloadducts.

5.6 Oxazonines

Nine membered heterocyclic rings represent a core structure to many natural products and bioactive molecules *(130)* and are also know to undergo unconventional transformations based on peripheral attacks *(131)*. In 2017, Zhao and coworkers synthetized benzofuran-fused oxazonines using the Pd-catalyzed [5+4] cycloaddition of azadienes with vinyl ethylene carbonates *(132)*. The use of diphosphine ligand (S,S)-**L25** provided the benzofuran-fused oxazonines in high enantioselectivities (up to 92% ee; Scheme 95A). Peripheral epoxidation of the double bond and subsequent treatment with $BF_3 \cdot Et_2O$ lead to the rearrangement of the nine-membered ring to a benzofuran-fused [6,6] polycyclic compound without erosion of selectivity (up to >20:1 dr and up to >99% ee; Scheme 95B). Exceptionally it was also observed that the nine-membered rings could undergo a ring contraction with the formation of four contiguous stereogenic centers and the migration of the tosyl group by simply refluxing the solution of the cycloadducts (Scheme 95C).

In 2020, Shibata and coworkers reported the Pd-catalyzed [5+4] cycloaddition based in the double decarboxylation of CF_3-decorated carbamates and vinyl ethylene carbonates to synthetize CF_3-benzo[*d*][1,3]oxazonines *(133)*. The use of Pd/(R)-**L15** catalytic system led to an efficient kinetic resolution (S factor up to 713; Scheme 96). Interestingly, the recovered (S)-starting carbamate could be converted in the enantiopure CF_3-benzo[*d*][1,3]oxazonines using $Pd(PPh_3)_4$ *(133)*.

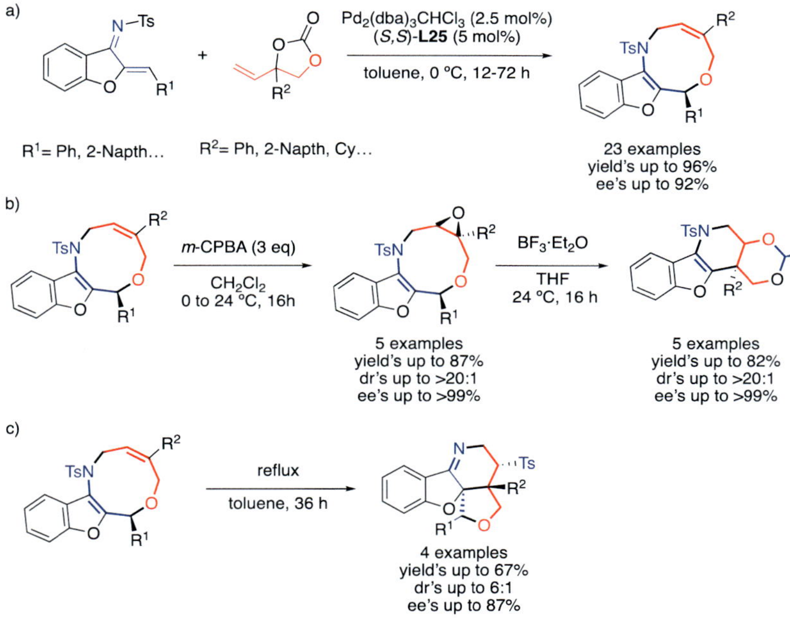

Scheme 95 (A) Pd-catalyzed [5+4] cycloaddition of azadienes with vinyl ethylene carbonates. (B) Peripheral epoxidation and subsequent rearrangement of the newly synthetized cycloadducts. (C) Ring contraction of the newly formed chiral oxazonines.

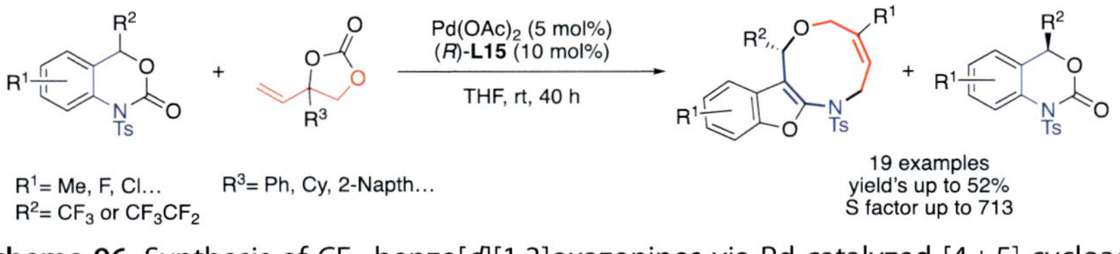

Scheme 96 Synthesis of CF_3-benzo[d][1,3]oxazonines via Pd-catalyzed [4+5] cycloaddition based in the double decarboxylation of between CF_3-decorated carbamates and vinyl ethylene carbonates.

5.7 Oxazecines

One of the most challenging ring formations is the 10-membered cycloadducts even though are present in many fields, especially drug discovery *(134)*. In 2018, Zhao and coworkers reported the Pd-catalyzed [6+4] cycloaddition reaction of vinyl oxetanes with azadienes to form benzofuran-fused and indene-fused oxazecines *(135)*. The use of spiroPHOX ligand (S,S)-**L56** provided the best enantioselectivities (up to >99% ee; Scheme 97A). Ring contraction of the newly formed oxazecines to tetrahydropyridines was performed to further prove the potential of these newly formed cyclo-adducts (Scheme 97B).

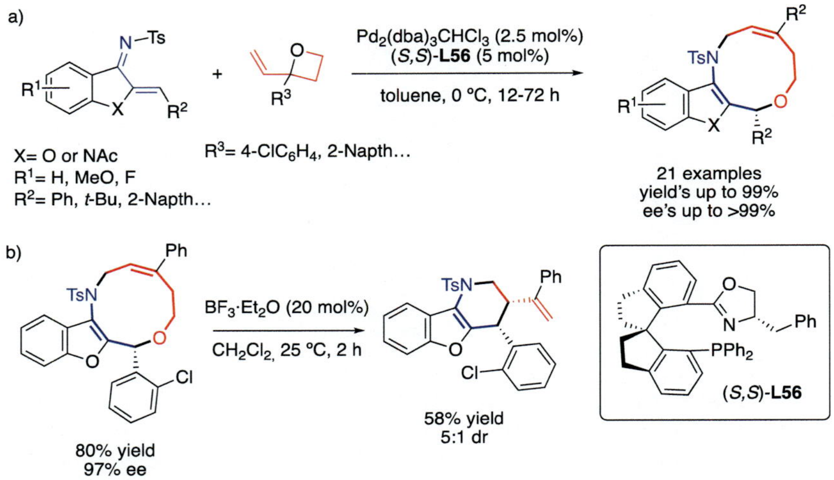

Scheme 97 (A) Pd-catalyzed [6+4] cycloaddition reaction of vinyl oxetanes with azadienes to form benzofuran-fused and indene-fused oxazecines. (B) Ring contraction of the newly formed oxazecines to tetrahydropyridines.

More recently, Xiao and coworkers developed a new Pd-catalyzed [8+2] cycloaddition which involved the *in situ* formation of a ketene synthon via Wolff rearrangement of α–diazoketones with vinyl carbamates *(136)*. Surprisingly, in the case of PMP-protected carbamate no decarboxylation took place, this new reactivity patter forms the 1,8-Pd-zwitterion, which with the concomitant addition to the in situ generated ketene yields the corresponding 10-membered cycloadduct *(136)*. The use of thioether-phosphoroamidite ligand (R)-**L57** provided the ten-membered rings in high yields and enantioselectivities (up to 94% ee; Scheme 98). Gram-scale synthesis and ring contraction reactions with retention of the optical purity were performed in order to prove the utility of the synthetic methodology *(136)*.

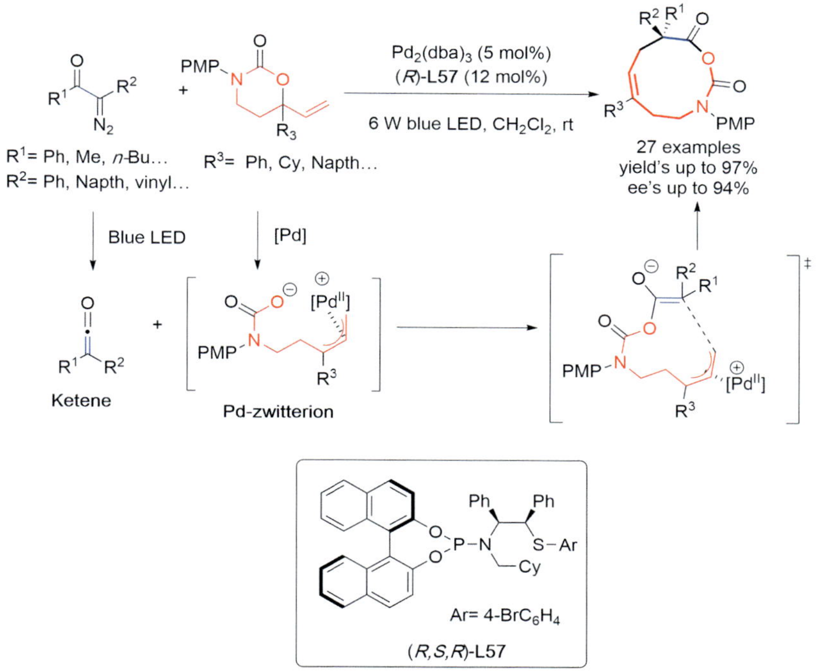

Metal-π-allyl mediated asymmetric cycloaddition reactions

Scheme 98 Pd-catalyzed [8 + 2] cycloaddition which involved the *in situ* formation of a ketene via Wolff rearrangement of α-diazoketones with vinyl carbamates.

6. Conclusions

The asymmetric metal-catalyzed cycloaddition reactions via π–allyl zwitterionic species continues to be one of the most important strategies for the construction of chiral carbo- and heterocycles. Nowadays a great variety of chiral cyclic structures with multiple stereogenic centers are therefore easily accessible with almost perfect diastereo- and enantioselectivity control.

Acknowledgments

We gratefully acknowledge financial support the Spanish Ministry of Economy and Competitiveness (PID2019-104904GB-I00), European Regional Development Fund (AEI/FEDER, UE) and the Catalan Government (2017SGR1472.

References

1. For reviews, see: (a) For reviews, see: He, J.; Ling, J.; Chiu, P. *Chem. Rev.* **2014**, *114*, 8037–8128; (b) Allen, D. W. B.; Lakeland, C. P.; Harrity, J. P. A. *Chem. A Eur. J.* **2017**, *23*, 13830–13857; (c) De, N.; Yoo, E. J. *ACS Catal.* **2018**, *8*, 48–58; (d) Liu, Y.; Oble, J.; Pradal, A.; Poli, G. *Eur. J. Inorg. Chem.* **2020**, 942–961; (e) Trost, B. M.; Mata, G. *Acc. Chem. Res.* **2020**, *53*, 1293–1305; (f) Wang, J.; Blaszczyk, S. A.; Li, X.; Tang, W. *Chem. Rev.* **2021**, *121*, 110–139; (g) Yang, C.; Yang, Z.-X.; Ding, C.-H.; Xu, B.; Hou, X.-L. *Chem. Rec.* **2021**, *21*, 1–14.
2. Pàmies, O.; Margalef, J.; Cañellas, S.; James, J.; Judge, E.; Guiry, P. J.; Moberg, C.; Bäckvall, J.-E.; Pfaltz, A.; Pericàs, M. A.; Diéguez, M. *Chem. Rev.* **2021**, *121*, 4373–4505.
3. See for instance: (a) See for instance: Guo, C.; Janssen-Müller, D.; Fleige, M.; Lerchen, A.; Daniliuc, C. G.; Glorius, F. *J. Am. Chem. Soc.* **2017**, *139*, 4443–4451; (b) Guo, C.; Fleige, M.; Janssen-Müller, D.; Daniliuc, C. G.; Glorius, F. *J. Am. Chem. Soc.* **2016**, *138*, 7840–7843; (c) Li, M.-M.; Wei, Y.; Liu, J.; Chen, H.-W.; Lu, L.-Q.; Xiao, W.-J. *J. Am. Chem. Soc.* **2017**, *139*, 14707–14713.
4. See for example: (a) See for example: Khan, A.; Yang, L.; Xu, J.; Jin, L. Y.; Zhang, Y. J. *Angew. Chem. Int. Ed.* **2014**, *53*, 11257–11260; (b) Cheng, Q.; Zhang, H.-J.; Yue, W.-J.; You, S.-L. *Chem* **2017**, *3*, 428–436; (c) Wu, Q.-Q.; Ding, C.-H.; Hou, X.-L. *Synlett* **2012**, *23*, 1035–1038.
5. Suo, J.-J.; Du, J.; Liu, Q.-R.; Chen, D.; Ding, C.-H.; Peng, Q.; Hou, X.-L. *Org. Lett.* **2017**, *19*, 6658–6661.
6. Wirth, T.; Kulicke, K. J.; Fragale, G. *J. Org. Chem.* **1996**, *61*, 2686–2689.
7. Swain, N. A.; Brown, R. C. D.; Bruton, G. A. *J. Org. Chem.* **2004**, *69*, 122–129.
8. Cheng, Q.; Zhang, H.-J.; Yue, W.-J.; You, S.-L. *Chem* **2017**, *3*, 428–436.
9. Cheng, Q.; Zhang, F.; Yue, C.; Guo, Y. L.; You, S.-L. *Angew. Chem. Int. Ed.* **2018**, *57*, 2134–2138.
10. Liu, K.; Khan, I.; Cheng, J.; Hsueh, Y.-J.; Zhang, Y.-J. *ACS Catal.* **2018**, *8*, 11600–11604.
11. Du, J.; Jiang, Y.-J.; Suo, J.-J.; Wu, W.-Q.; Liu, X.-Y.; Chen, D.; Ding, C.-H.; Wei, Y.; Hou, X.-L. *Chem. Commun.* **2018**, *54*, 13143–13146.
12. Khan, I.; Zhao, C.; Zhang, Y.-J. *Chem. Commun.* **2018**, *54*, 4708–4711.
13. Wang, W.-Y.; Wu, J.-Y.; Liu, Q.-R.; Liu, X.-Y.; Ding, C.-H.; Hou, X.-L. *Org. Lett.* **2018**, *20*, 4773–4776.
14. Huang, K.-X.; Xie, M.-S.; Wang, D.-C.; Sang, J.-W.; Qu, G.-R.; Guo, H.-M. *Chem. Commun.* **2019**, *55*, 13550–13553.
15. Wang, J.; Zhao, L.; Rong, Q.; Lv, C.; Lu, Y.; Pan, X.; Zhao, L.; Hu, L. *Org. Lett.* **2020**, 5833–5838.
16. Zhang, H.; Gao, X.; Jiang, F.; Shi, W.; Wang, W.; Wu, Y.; Zhang, C.; Shi, X.; Guo, H. *Eur. J. Org. Chem.* **2020**, *30*, 4301–4804.
17. Zhao, C.; Khan, I.; Zhang, Y. J. *Chem. Commun.* **2020**, *56*, 12431–12434.
18. Xiao, J.-A.; Li, Y.-C.; Luo, Z.-J.; Cheng, X.-L.; Deng, Z.-X.; Chen, W.-Q.; Su, W.; Yang, H. *J. Org. Chem.* **2019**, *84*, 2297–2306.
19. Jie, L.; Li, M.-M.; Qu, B.-L.; Lu, L.-Q.; Xiao, W.-J. *Chem. Commun.* **2019**, *55*, 2031–2034.
20. Mondal, M.; Panda, M.; McKee, V.; Kerrigan, N. J. *J. Org. Chem.* **2019**, *84*, 11983–11991.
21. Trost, B. M.; Wang, Y.; Hung, C.-I. *Nat. Chem.* **2020**, *12*, 294–301.
22. See for instance: (a) See for instance: Ozoe, Y.; Asahi, M.; Ozoe, F.; Nakahira, K.; Mita, T. *Biochem. Biophys. Res. Commun.* **2010**, *391*, 744–749; (b) Murphy, M.; Cavalleri, D.; Seewald, W.; Drake, J.; Nanchen, S. *Parasit. Vectors* **2017**, *10*, 541–549; (c) El Qacemil, M. C.; Toueg, J. Y.; Renold, J. C.; Pitter-Na, P. (Syngenta Participations AG), WO 2011/101229 A1, 2011.

23. Trost, B. M.; Guillaume, M. *Angew. Chem. Int. Ed.* **2018**, *57*, 12333–12337.
24. Mao, B.; Liu, H.; Yan, Z.; Xu, Y.; Xu, J.; Wang, W.; Wu, Y.; Guo, H. *Angew. Chem. Int. Ed.* **2020**, *59*, 11316–11320.
25. Du, J.; Gua, Y.-D.; Jiang, Y.-J.; Huang, S.; Chen, D.; Ding, C.-H.; Hou, X.-L. *Org. Lett.* **2020**, *22*, 5375–5379.
26. See for instance: (a) See for instance: Wang, Y.-G.; Takeyama, R.; Kobayashi, Y. *Angew. Chem. Int. Ed.* **2006**, *45*, 3320–3323; (b) Nakamura, T.; Shirokawa, S.; Hosokawa, S.; Nakazaki, A.; Kobayashi, S. *Org. Lett.* **2006**, *8*, 677–679; (c) Yao, Y.-S.; Liu, J.-L.; Xi, J.; Miu, B.; Liu, G.-S.; Wang, S.; Meng, L.; Yao, Z.-J. *Chem. A Eur. J.* **2011**, *17*, 10462–10469.
27. Xu, H.; Khan, S.; Li, H.; Wu, X.; Zhang, Y. J. *Org. Lett.* **2019**, *21*, 214–217.
28. Shiina, I.; Nakata, K. Medium-sized lactones. In *Natural Lactones and Lactams*; Janecki, T., Ed.; 2013.
29. Singha, S.; Patra, T.; Daniliuc, C. G.; Glorious, F. *J. Am. Chem. Soc.* **2018**, *140*, 3551–3554.
30. Singha, S.; Serrano, E.; Mondal, S.; Daniliuc, C. G.; Glorious, F. *Nat. Catal.* **2020**, *3*, 48–54.
31. Wei, Y.; Liu, S.; Li, M.-M.; Li, Y.; Lan, Y.; Lu, L.-Q.; Xiao, W.-J. *J. Am. Chem. Soc.* **2019**, *141*, 133–137.
32. Li, M. M.; Xiong, Q.; Qu, B.-L.; Xiao, Y.-Q.; Lan, Y.; Lu, L.-Q.; Xiao, W.-J. *Angew. Chem. Int. Ed.* **2020**, *59*, 17429–17434.
33. Gao, X.; Xia, M.; Yuan, C.; Zhou, L.; Sun, W.; Li, C.; Wu, B.; Zhu, D.; Zhang, C.; Zheng, B.; Wang, D.; Guo, H. *ACS Catal.* **2019**, *9*, 1645–1654.
34. See for instance: (a) See for instance: Pegoraro, S.; Lang, M.; Dreker, T.; Kraus, J.; Hamm, S.; Meere, C.; Freurle, J.; Tasler, S.; Prütting, S.; Kura, Z.; Visan, V.; Grissmer, S. *Bioorg. Med. Chem. Lett.* **2009**, *19*, 2299–2304; (b) Torres-Marquez, E.; Sinnett-Smith, J.; Ghua, S.; Kui, R.; Waldron, T.; Rey, O.; Rozegurt, E. *Biochem. Biophys. Res. Commun.* **2010**, *391*, 63–68; (c) Zhang, B.; Bao, M.; Zeng, C.; Zhong, X.; Ni, L.; Zeng, Y.; Cai, X. *Org. Lett.* **2014**, *16*, 6400–6403.
35. Trost, B. M.; Zuo, Z. *Angew. Chem. Int. Ed.* **2020**, *59*, 1243–1247.
36. An, X.-T.; Du, J.-Y.; Jia, Z.-L.; Zhang, Q.; Yu, K.-Y.; Zhang, Y.-H.; Zhao, X.-H.; Fang, R.; Fan, C.-A. *Chem. A Eur. J.* **2020**, *26*, 3803–3809.
37. Examples: (a) Examples: Vitaku, E.; Smith, D. T.; Njaderson, J. T. *J. Med. Chem.* **2014**, *57*, 10257–11074; (b) Taylor, R.; MacCoss, D. M.; Lawson, A. D. G. *J. Med. Chem.* **2014**, *57*, 5845–5859; (c) Aldeghi, M.; Malhotra, M.; Selwood, D. L.; Chan, A. W. E. *Chem. Biol. Drug Des.* **2014**, *83*, 450–461; (d) Martins, P.; Jesus, J.; Santos, S.; Raposo, L. R.; Roma-Rodrigues, C.; Baptista, P. V.; Fernandes, A. R. *Molecules* **2015**, *20*, 16852–16891.
38. See for instance: (a) See for instance: Butler, D. C. D.; Inman, G. A.; Alper, H. *J. Org. Chem.* **2000**, *65*, 5887–5890; (b) Aoyagi, K.; Nakamura, H.; Yamamoto, Y. *J. Org. Chem.* **2002**, *67*, 5977–5980; (c) Trost, B. M.; Fandrick, D. R. *Org. Lett.* **2005**, *7*, 823–826; (d) Fontana, F.; Chen, C. C.; Aggarwal, V. K. *Org. Lett.* **2011**, *13*, 3454–3457; (e) Lowe, M. A.; Ostovar, M.; Ferrini, S.; Chen, C. C.; Lawrence, P. G.; Fontana, F.; Calabrese, A. A.; Aggarwal, V. K. *Am. Ethnol.* **2011**, *123*, 6494–6498; (f) Feng, J.-J.; Lin, T.-Y.; Zhu, C.-Z.; Wang, H.; Wu, H.-H.; Zhang, J. *J. Am. Chem. Soc.* **2016**, *138*, 2178–2181; (g) Xu, C.-F.; Zheng, B.-H.; Suo, J.-J.; Ding, C.-H.; Hou, X.-L. *Angew. Chem. Int. Ed.* **2015**, *54*, 1604–1607; (h) Hashimoto, T.; Takino, K.; Hato, K.; Maruoka, K. *Angew. Chem. Int. Ed.* **2016**, *55*, 8081–8085.
39. Næsborg, L.; Tur, F.; Meazza, M.; Blom, J.; Halskov, K. S.; Jørgensen, K. A. *Chem. A Eur. J.* **2017**, *23*, 268–272.

40. Zhang, J.-Q.; Tong, F.; Sun, B.-B.; Fan, W.-T.; Chen, J.-B.; Hu, D.; Wang, X.-W. *J. Org. Chem.* **2018**, *83*, 2882–2891.
41. Suo, J.-J.; Liu, W.; Ding, C.-H.; Hou, X.-L. Chem. Asian J. 2018, 13, 959–963.
42. Vetica, F.; Baileya, S. J.; Kumara, M.; Mahajana, S.; Essen, C. V.; Rissanen, K.; Endersa, D. *Synthesis* **2020**, *52*, 2038–2044.
43. Wang, Q.; Wang, C.; Shi, W.; Xiao, Y.; Guo, H. *Org. Biomol. Chem.* **2018**, *16*, 4881–4887.
44. Ling, J.; Laugeois, M.; Ratovelomanana-Vidal, V.; Vitale, M. R. *Synlett* **2018**, *29*, 2288–2292.
45. Huang, X.-B.; Li, X.-J.; Li, T.-T.; Chen, B.; Chu, W.-D.; He, L.; Liu, Q.-Z. *Org. Lett.* **2019**, *21*, 1713–1716.
46. Song, X.; Gu, M.; Chen, X.; Xu, L.; Ni, Q. *Asian J. Org. Chem.* **2019**, *8*, 2180–2183.
47. Trost, B. M.; Shinde, A. H.; Wang, Y.; Zui, Z.; Min, C. *ACS Catal.* **2020**, *10*, 1969–1975.
48. See for example: (a) See for examples: Migliori, G. B.; Dheda, K.; Centis, R.; Mwaba, P.; Bates, M.; O'Grady, J.; Hoelscher, M.; Zumla, A. *Trop. Med. Int. Health* **2010**, *15*, 1052–1066; (b) Drawz, S. M.; Bonomo, R. A. *Clin. Microbiol. Rev.* **2010**, *23*, 160–201; (c) Ersmark, K.; Del Valle, J. R.; Hanessian, S. *Angew. Chem. Int. Ed.* **2008**, *47*, 1202–1223; (d) Sayago, F. J.; Laborda, P.; Calaza, M. I.; Jimeńez, A. I.; Cativiela, C. *Eur. J. Org. Chem.* **2011**, *16*, 3074–3081.
49. Lu, C.-H.; Darvishi, S.; Khakyzadeh, V.; Li, C. *Chin. Chem. Lett.* **2021**, *31*, 405–507.
50. See for instance: (a) See for instance: Liu, W.-J.; Chen, X.-H.; Gong, L.-Z. *Org. Lett.* **2008**, *10*, 5357–5360; (b) Li, Q.-H.; Wei, L.; Chen, X.; Wang, C.-J. *Chem. Commun.* **2013**, *49*, 6277–6279; (c) Ohmatsu, K.; Kawai, S.; Imagawa, N.; Ooi, T. *ACS Catal.* **2014**, *4*, 4304–4306; (d) Armstrong, R. J.; D'Ascenzio, M.; Smith, M. D. *Synlett* **2016**, *27*, 6–10; (e) Sawatzky, E.; Drakopoulos, A.; Rolz, M.; Sotriffer, C.; Engels, B.; Decker, M. *Beilstein J. Org. Chem.* **2016**, *12*, 2280–2292; (f) Trost, B. M.; Gnanamani, E.; Hung, C.-I. *Angew. Chem. Int. Ed.* **2017**, *56*, 10451–10456; (g) Mukhopadhyay, S.; Pan, S. C. *Chem. Commun.* **2018**, *54*, 964–967.
51. Spielmann, K.; Lee, A. V. D.; Figueiredo, R. M.; Campagne, J. M. *Org. Lett.* **2018**, *20*, 1444–1447.
52. Spielmann, K.; Tosi, E.; Lebrun, A.; Niel, G.; Lee, A. V. D.; Figueiredo, R. M.; Campagne, J. M. *Tetrahedron* **2019**, *74*, 6487–6511.
53. See for instance: (a) See for example: Rauckman, B. S.; Tidwell, M. Y.; Johnson, J. V.; Roth, B. *J. Med. Chem.* **1989**, *32*, 1927–1935; (b) Katritzky, A. R.; Rachwal, S.; Rachwal, B. *Tetrahedron* **1996**, *52*, 15031–15070; (c) Barton, D. H.; Nakanishi, K.; Cohn, O. M. *Comprehensive Natural Products Chemistry*; Elsevier: Oxford, 1999; (d) Sridharan, V. P.; Suryavanshi, A.; Menendez, J. C. *Chem. Rev.* **2011**, *111*, 7157–7259.
54. See for instance: (a) See for example: Wang, C.; Tunge, J. A. *J. Am. Chem. Soc.* **2008**, *120*, 8118–8119; (b) Wei, Y.; Lu, L. Q.; Li, T. R.; Wang, Q.; Ciao, W.-J.; Alper, H. *Angew. Chem. Int. Ed.* **2016**, *55*, 2200–2204; (c) Leth, L. A.; Glaus, F.; Meazza, M.; Fu, L.; Thøgersen, M. K.; Bitsch, E. A.; Jørgensen, K. A. *Angew. Chem. Int. Ed.* **2016**, *55*, 15272–15276.
55. Mei, G.-J.; Li, D.; Xhou, G.-X.; Shi, Q.; Cao, Z.; Shi, F. *Chem. Commun.* **2017**, *53*, 10030–10033.
56. Zhao, H.-W.; Feng, N.-N.; Guo, J.-M.; Du, J.; Ding, W.-Q.; Wang, L.-R.; Dong, X.-Q. *J. Org. Chem.* **2018**, *83*, 9291–9299.
57. Suo, J.-J.; Du, J.; Jiang, Y.-J.; Chen, D.; Ding, C.-H.; Hou, X.-L. *Chin. Chem. Lett.* **2019**, *30*, 1512–1514.
58. Guo, J.-M.; Fan, X.-Z.; Wu, H.-H.; Tang, Z.; Bi, X.-F.; Xhang, H.; Cai, L.-Y.; Xhao, H.-W.; Zhong, Q.-D. *J. Org. Chem.* **2021**, *86*, 1712–1720.

59. See for instance: (a) See for example: Seo, H. N.; Choi, J. Y.; Choe, Y. J.; Kim, Y.; Rhim, H.; Lee, S. H.; Kim, J.; Joo, D. J.; Lee, J. Y. *Bioorg. Med. Chem. Lett.* **2007**, *17*, 5740–5743; (b) Michael, J. P. *Nat. Prod. Rep.* **2007**, *24*, 223–246; (c) Liu, L. T.; Yuan, T. T.; Liu, H. H.; Chen, S. F.; Wu, Y. T. *Bioorg. Med. Chem. Lett.* **2007**, *17*, 6373–6377; (d) Gundla, R.; Kazemi, R.; Sanam, R.; Muttineni, R.; Sarma, J. A.; Dayam, R.; Neamati, N. *J. Med. Chem.* **2008**, *51*, 3367–3377; (e) Da Silva, J. F.; Walters, M.; Al-Damluji, S.; Ganellin, C. R. *Bioorg. Med. Chem.* **2008**, *16*, 7254–7263; (f) Hasegawa, H.; Muraoka, M.; Matsui, K.; Kojima, A. *Bioorg. Med. Chem. Lett.* **2003**, *13*, 3471–3476; (h) Corbett, J. W.; Ko, S. S.; Rodgers, J. D.; Gearhart, L. A.; Magnus, N. A.; Bacheler, L. T.; Diamond, S.; Jeffrey, S.; Klabe, R. M.; Cordova, B. C.; Garber, S.; Logue, K.; Trainor, G. L.; Anderson, P. S.; Erickson-Viitanen, S. K. *J. Med. Chem.* **2000**, *43*, 2019–2030.

60. Wang, Q.; Li, Y.; Wu, Y.; Wang, Q.; Shi, W.; Yuan, C.; Zhou, L.; Xiao, Y.; Guo, H. *Org. Lett.* **2018**, *20*, 2880–2883.

61. Mun, D.; Kim, E.; Kim, S.-G. *Synthesis* **2019**, *51*, 2359–2370.

62. See for instance: (a) See for example: Chang, C.-F.; Hsu, Y.-L.; Lee, C.-Y.; Wu, C.-H.; Wu, Y.-C.; Chuang, T.-H. *J. Mol. Sci.* **2015**, *16*, 3980–3989; (b) Kshirsagar, U. A. *Biomol. Chem.* **2015**, *13*, 9336–9352; (c) Yang, S.; Li, X.; Hu, F.; Li, Y.; Yang, Y.; Yan, J.; Kuang, C.; Yang, Q. *J. Med. Chem.* **2013**, *56*, 8321–8331; (d) Chen, M.; Gan, L.; Lin, S.; Wang, X.; Li, L.; Li, Y.; Zhu, C.; Wang, Y.; Jiang, B.; Jiang, J.; Yang, Y.; Shi, J. *J. Nat. Prod.* **2012**, *75*, 1167–1176; (e) Jao, C. W.; Lin, W. C.; Wu, Y. T.; Wu, P. L. *J. Nat. Prod.* **2008**, *71*, 1275–1279.

63. Mei, G.-J.; Bian, C.-Y.; Li, G.-H.; Xu, S.-L.; Zheng, W.-Q.; Shi, F. *Org. Lett.* **2017**, *19* (12), 3219–3222.

64. See for instance: (a) See for instance: Nájera, C.; Sansano, J. M.; Yus, M. *Org. Biomol. Chem.* **2015**, *13*, 8596–8636; (b) Chen, Z.; Wu, J. *Org. Lett.* **2010**, *12* (21), 4856–4859.

65. Mao, B.; Xu, Y.; Chen, Y.; Dong, J.; Zhang, J.; Gu, K.; Xheng, B.; Guo, H. *Org. Lett.* **2019**, *21*, 4424–4427.

66. Mao, B.; Zhang, J.; Xu, Y.; Yan, Z.; Wang, W.; Wu, Y.; Sun, C.; Zheng, B.; Guo, H. *Chem. Commun.* **2019**, *55*, 12841–12844.

67. See for instance: (a) See for example: Majumdar, K. C.; Chattopadhyay, S. K. *Heterocycles in Natural Product Synthesis*, 1st ed.; Wiley-VCH, Weinheim: Germany, 2011; (b) Vitaku, E.; Smith, D. T.; Njardarson, J. T. *J. Med. Chem.* **2014**, *57*, 10257–10274; (c) Horton, D. A.; Bourne, G. T.; Smythe, M. L. *Chem. Rev.* **2003**, *103*, 893–930; (d) Boente, J. M.; Campello, M. J.; Castedo, L.; Dominguez, D.; Saa, J. M.; Suau, R.; Vidal, M. C. *Tetrahedron Lett.* **1983**, *24*, 4481–4484; (e) Chang, Y.; Meng, F.-C.; Wang, R.; Wang, C. M.; Lu, X.-Y.; Zhang, Q.-W. In *In Studies in Natural Products Chemistry*; Rahman, A., Ed.; Vol. 53; Elsevier, 2017; pp. 339–373; (f) Marco-Contelles, J.; do Carmo Carreiras, M.; Rodriguez, C.; Villarroya, M.; Garcia, A. G. *Chem. Rev.* **2006**, *106*, 116–133; (g) Lavaud, C.; Massiot, G. The iboga alkaloids. In *Progress in the Chemistry of Organic Natural Products*; Kinghorn, A., Falk, H., Gibbons, S., Kobayashi, J., Eds.; Vol. 105; Springer: Cham, 2017; pp. 89–136; (h) Scheiter, M.; Bulitta, B.; van Ham, M.; Klawonn, F.; Koenig, S.; Jaensch, L. *Front. Immunol.* **2013**, *4*, 66–76; (i) Hou, F. F.; Zhang, X.; Zhang, G. H.; Xie, D.; Chen, P. Y.; Zhang, W. R.; Jiang, J. P.; Liang, M.; Wang, G. B.; Liu, Z. R.; Geng, R. W. *N. Engl. J. Med.* **2006**, *354*, 131–140.

68. Chen, Z.-C.; Chen, Z.; Yang, Z.-H.; Guo, L.; Du, W.; Chen, Y.-C. *Angew. Chem. Int. Ed.* **2019**, *58*, 15021–15025.

69. Liu, Y.-Z.; Wang, Z.; Huang, Z.; Zheng, X.; Yang, W.-L.; Deng, W.-P. *Angew. Chem. Int. Ed.* **2020**, *59*, 1238–1242.

70. Kumari, P.; Liu, W.; Wang, C.-J.; Dai, J.; Wang, M.-X.; Yang, Q.-Q.; Deng, Y.-H.; Shao, Z. *Chin. J. Chem.* **2020**, *38*, 151–157.

71. Yang, W.-L.; Huang, Z.; Liu, Y.-Z.; Yu, X.; Deng, W.-P. *Chin. J. Chem.* **2020**, *38*, 1571–1574.
72. Liu, Y.-Z.; Wang, Z.; Huang, Z.; Yang, W.-L.; Deng, W.-P. *Org. Lett.* **2021**, *23*, 948–952.
73. See for instance: (a) See for example: Meyers, A. I.; Brengel, G. P. *Chem. Commun.* **1997**, 1–8; (b) Newhause, B.; Allen, S.; Fauber, B.; Anderson, A. S.; Eary, C.-T.; Hansen, J. D.; Schiro, J.; Gaudino, J. J.; Laird, E.; Chantry, D.; Eberhardt, C.; Burguess, L. E. *Bioorg. Med. Chem. Lett.* **2005**, *14*, 5537–5542; (c) Kang, G.; Yamagami, M.; Vellalath, S.; Romo, D. *Angew. Chem. Int. Ed.* **2018**, *57*, 6527–6531; (d) Fischer, C.; Zultanski, S. L.; Zhou, H.; Methot, J. L.; Shah, S.; Hayashi, I.; Hughes, B. L.; Moxham, C. M.; Bays, N. W.; Smotrov, N.; Hill, A. D.; Pan, B.-S.; Wu, Z.; Moy, L. Y.; Tanga, F.; Kenific, C.; Cruz, J. C.; Walker, D.; Bouthillette, M.; Nikov, G. N.; Deshmukh, S. V.; Jeliazkova-Mecheva, V. V.; Diaz, D.; Munoz, B.; Shearman, M. S.; Michener, M. S.; Cook, J. J. *Bioorg.Med. Chem. Lett.* **2015**, *25*, 3488–3494; (e) Kumar Boominathan, S. S.; Reddy, M. M.; Hou, R.-J.; Chen, H.-F.; Wang, J.-J. *Org. Biomol. Chem.* **2017**, *15*, 1872–1875; (f) Patel, R. N. *Adv. Synth. Catal.* **2001**, *343*, 527–546; (g) Zhou, B.; Li, L.; Zhu, X.-Q.; Yan, J.-Z.; Guo, Y.-L.; Ye, L.-W. *Angew. Chem. Int. Ed.* **2017**, *56*, 4015–4019.
74. Jin, J.-H.; Wang, H.; Yang, Z.-T.; Yang, W.-L.; Tang, W.; Deng, W.-P. *Org. Lett.* **2018**, *20*, 104–107.
75. Wang, Y.-N.; Xiong, Q.; Lu, L.-Q.; Zhang, Q.-L.; Wang, Y.; Lan, Y.; Xiao, W.-J. *Angew. Chem. Int. Ed.* **2019**, *58*, 11013–11017.
76. Gao, J.; Xhang, J.; Fang, S.; Feng, J.; Lu, T.; Du, D. *Org. Lett.* **2020**, *19*, 7725–7729.
77. Than, F.; Yang, W.-L.; Ni, T.; Zhang, J.; Deng, W.-P. *Sci. China Chem.* **2021**, *64*, 34–40.
78. See for instance: (a) See for example: Moreau, F.; Florentin, D.; Marquet, A. *Tetrahedron* **2000**, *56*, 285–293; (b) Lien, E. J.; Hussain, M.; Golden, M. P. *J. Med. Chem.* **1970**, *13*, 623–626; (c) Kazmierski, W. M.; Furfine, E.; Gray-Nunez, Y.; Spaltenstein, A.; Wright, L. *Bioorg. Med. Chem. Lett.* **2004**, *14*, 5685–5687; (d) Li, J. J.; Sutton, J. C.; Nirschl, A.; Zou, Y.; Wang, H.; Sun, C.; Pi, Z.; Johnson, R.; Krystek, S. R.; Seethala, R. *J. Med. Chem.* **2007**, *50*, 3015–3025.
79. Lu, Y.-N.; Lan, J.-P.; Mao, Y.-J.; Wang, Y.-W.; Mei, G.-J.; Shi, F. *Chem. Commun.* **2018**, *54*, 13527–13530.
80. Khan, I.; Shah, B. H.; Zhao, C.; Xu, F.; Zhang, Y. J. *Org. Lett.* **2019**, *21*, 9452–9456.
81. Hang, Q.-Q.; Liu, S.-J.; Yu, L.; Sun, T.-T.; Zhang, Y.-C.; Mei, G.-J.; Shi, F. *Chin. J. Chem.* **2020**, *38*, 1612–1618.
82. See for instance: (a) See for instance: Melian, E. B.; Goa, K. L. *Drugs* **2002**, *62*, 107–133; (b) Reino, J. L.; Duran-Patron, R.; Segura, I.; Hernandez-Galan, R.; Riese, H. H.; Collado, I. G. *J. Nat. Prod.* **2003**, *66*, 344–349; (c) Kinnel, R. B.; Gehrken, H.-P.; Scheuer, P. J. *J. Am. Chem. Soc.* **1993**, *115*, 3376–3377; (d) Kinnel, R. B.; Gehrken, H.-P.; Swali, R.; Skoropowski, G.; Scheuer, P. J. *J. Org. Chem.* **1998**, *63*, 3281–3287.
83. See for instance: (a) See For example: Trost, B. M.; Morris, P. J. *Angew. Chem. Int. Ed.* **2011**, *50*, 6167–6170; (b) Mei, L.-Y.; Wei, Y.; Xu, Q.; Shi, M. *Organometallics* **2012**, *31*, 7591–7955; (c) Trost, B. M.; Morris, P. J.; Sprague, S. J. *J. Am. Chem. Soc.* **2012**, *134*, 17823–17831; (d) Liu, Z.-S.; Li, W.-K.; Kang, T.-R.; He, L.; Liu, Q.-Z. *Org. Lett.* **2015**, *17*, 150–153; (e) Xie, M.-S.; Wang, Y.; Li, J.-P.; Du, C.; Zhang, Y.-Y.; Hao, E.-J.; Zhang, Y.-M.; Qu, G.-R.; Guo, H.-M. *Chem. Commun.* **2015**, *51*, 12451–12454; (f) Ma, C.; Huang, Y.; Zhao, Y. *ACS Catal.* **2016**, *6*, 6408–6412; (j) Halskov, K. S.; Naesborg, L.; Tur, F.; Jøergensen, K. A. *Org. Lett.* **2016**, *18*, 2220–2223; (h) Laugeois, M.; Ratovelomanana-Vidal, V.; Michelet, V.; Vitale, M. R. *Chem. Commun.* **2016**, *52*, 5332–5335.

84. Kamlar, M.; Franc, M.; Císařová, I.; Gyepes, R.; Veselý, J. *Chem. Commun.* **2019**, *55*, 3829–3832.
85. Meazza, M.; Kamlar, M.; Jašíková, L.; Formánek, B.; Mazzanati, A.; Roithová, J.; Veselý, J.; Rios, R. *Chem. Sci.* **2018**, *9*, 6368–6373.
86. Zhou, Q.; Chen, B.; Huang, X.-B.; Zeng, Y.-L.; Chu, W.-D.; He, L.; Liu, Q.-Z. *Org. Chem. Front.* **2019**, *6*, 1891–1894.
87. Wan, Q.; Chen, L.; Li, S.; Kang, Q.; Yuan, Y.; Du, Y. *Org. Lett.* **2020**, *22*, 9539–9544.
88. Liu, K.; Yang, J.; Li, X. *Org. Lett.* **2021**, *23*, 826–831.
89. Zhang, J.; Ma, K.; Chen, H.; Wang, K.; Xiong, W.; Bao, L.; Liu, H. *J. Antibiot.* **2017**, *70*, 915–917.
90. See for instance: (a) See for example: Samago, S.; Nakamura, E. *Org. React.* **2002**, *61*, 1–217; (b) Trost, B. M. *Angew. Chem. Int. Ed.* **1986**, *25*, 1–20.
91. Trost, B. M.; Zhang, L.; Lam, T. M. *Org. Lett.* **2018**, *20*, 3938–3942.
92. Trost, B. M.; Wang, Y. *Angew. Chem. Int. Ed.* **2018**, *57*, 11025–11029.
93. Maring, C. J.; Chen, Y.; Degoey, D. A.; Giranda, V. L.; Grampovnik, D. J.; Gu, Y. G.; Kati, W. M.; Kempf, D. J.; Kennedy, A.; Krueger, A. C.; Lin, Z.; Madigan, D. L.; Muchmore, S. W.; Sham, H. L.; Stewart, K. D.; Stoll, V. S.; Sun, M.; Wang, G.T.; Wang, S.; Yeung, M.C.; Zhao, C. Five-membered carbocyclic and heterocyclic inhibitors of neuraminidases U.S. Patent 6518305B1, 2003.
94. Trost, B. M.; Zell, D.; Christoph, H.; Mata, G.; Maruniak, A. *Angew. Chem. Int. Ed.* **2018**, *57*, 12916–12920.
95. Trost, B. M.; Jiao, Z.; Hung, C.-I. *Angew. Chem. Int. Ed.* **2019**, *58*, 15154–15158.
96. Trost, B. M.; Jiao, Z.; Liu, Y.; Min, C.; Hung, C.-I. *J. Am. Chem. Soc.* **2020**, *142*, 18628–18636.
97. Kumamoto, H.; Fukano, M.; Imoto, S.; Kohgo, S.; Odanaka, Y.; Amano, M.; Kuwata-Higashi, N.; Mitsuya, H.; Haraguchi, K.; Fukuhara, K. *Nucleosides Nucleotides Nucleic Acids* **2017**, *36*, 463–473.
98. Singh, U. S.; Chu, C. K. *Nucleosides Nucleotides Nucleic Acids* **2020**, *39*, 52–68.
99. Trost, B. M.; Zuo, Z.; Wang, Y. *Org. Lett.* **2021**, *23*, 979–983.
100. See for instance: (a) See for example: Harmata, M. *Chem. Commun.* **2010**, *46*, 8886–8903; (b) Harmata, M. *Chem. Commun.* **2010**, *46*, 8904–8922; (c) Lohse, A. G.; Hsung, R. P. *Chem. A Eur. J.* **2011**, *17*, 3812–3822; (d) Li, H.; Wu, J. *Synthesis* **2014**, *47*, 22–33.
101. Zheng, Y.; Qin, Z.; Zi, W. *J. Am. Chem. Soc.* **2021**, *142*, 1038–1045.
102. See for instance: (a) See for example: Jaroszewski, J. W.; Olafsdottir, E. S. *Tetrahedron Lett.* **1986**, *27*, 5297–5300; (b) Coburn, R. A.; Long, L., Jr. *J. Org. Chem.* **1966**, *31*, 4312–4315.
103. See for instance: (a) See for example: Vince, R.; Hua, M.; Brownell, J.; Daluge, S.; Lee, F.; Shannon, W. M.; Lavelle, G. C.; Qualls, J.; Weislow, O. S.; Kiser, R.; Canonico, P. G.; Schultz, R. H.; Narayanan, V. L.; Mayo, J. G.; Shoemaker, R. H.; Boyd, M. R. *Biochem. Biophys. Res. Commun.* **1988**, *156*, 1046–1053; (b) Daluge, S. M.; Good, S. S.; Faletto, M. B.; Miller, W. H.; Marty, H.; Clair, S. T.; Boone, L. R.; Tisdale, M.; Parry, N. R.; Reardon, J. E.; Dornsife, R. E.; Averett, D. R.; Krenitsky, T. A. *Antimicrob. Agents Chemother.* **1997**, *41*, 1082–1093.
104. Ding, W.-P.; Zhang, G.-P.; Jiang, Y.-J.; Du, J.; Liu, X.-Y.; Chen, D.; Ding, C.-H.; Deng, Q.-H.; Hou, X.-L. *Org. Lett.* **2019**, *21*, 6805–6810.
105. See for instance: (a) See for instance: Michalson, E. T.; Szmuszkovicz, J. *Prog. Drug Res.* **1989**, *22*, 135–149; (b) Fukuta, Y.; Mita, T.; Fukuda, N.; Kanai, M.; Shibasaki, M. *J. Am. Chem. Soc.* **2006**, *128*, 6312–6313; (c) Nakatani, N.; Inatani, R. *Agric. Biol. Chem.* **1983**, *47*, 353–358; (d) Lin, H.; Danishefsky, S. J. *Angew. Chem. Int. Ed.* **2003**, *42*, 36–51; (e) Mansfield, D.; Coqueron, P.-Y.; Rieck, H.; Desbordes, P.; Villier, A.; Grosjean-Cournoyer, M.-C.; Genix, P. 2-pyridinylcycloal- kylcarboxamide derivatives as fungicides. U.S. Pat. US 8071629 B2 20111206 (2011).

106. See for instance: (a) See for example: Lautens, M.; Klute, W.; Tam, W. *Chem. Rev.* **1996**, *96*, 49–92; (b) Fruhauf, H.-W. *Chem. Rev.* **1997**, *97*, 523–596; (c) Moyano, A.; Rios, R. *Chem. Rev.* **2011**, *111*, 4703–4832.

107. Jia, Z.-L.; An, X.-T.; Deng, Y.-H.; Pang, L. H.; Liu, C.-F.; Meng, L.-L.; Xue, J.-K.; Zhao, X.-H.; Fan, C.-H. *Org. Lett.* **2021**, *23*, 745–750.

108. Nair, V.; Deepthi, A.; Ashok, D.; Raveendran, A. E.; Paul, R. R. *Tetrahedron* **2014**, *70*, 3085–3105.

109. Trost, B. M.; Jiao, Z. *J. Am. Chem. Soc.* **2020**, *142*, 21645–21650.

110. Wang, F.; Dong, Z.-J.; Liu, J.-K. *Z. Naturforsch* **2007**, *62*, 1329–1332.

111. Caloprisco, E.; Fourneron, J.-D.; Faure, R.; Demarne, F.-E. *J. Agric. Food Chem.* **2002**, *50*, 78–80.

112. (a) Sperry, J.; Wilson, Z. E.; Rathwell, D. C. K.; Brimble, M. A. *Nat. Prod. Rep.* **2010**, *27*, 1117–1137; (b) Bartoli, A.; Rodier, F.; Commeiras, L.; Parrain, J.-L.; Chouraqui, G. *Nat. Prod. Rep.* **2011**, *28*, 763–782; (c) Müller, G.; Berkenbosch, T.; Benningshof, J. C. J.; Stumpfe, D.; Bajorath, J. *Chem. A Eur. J.* **2017**, *23*, 703–710.

113. Yang, L.-C.; Tan, Z. Y.; Rong, Z.-Q.; Liu, R.; Wang, Y.-N.; Zhao, Y. *Angew. Chem. Int. Ed.* **2018**, *57*, 7860–7864.

114. Stempel, E.; Gaich, T. *Acc. Chem. Res.* **2016**, *49* (11), 2390–2402.

115. Zheng, X.; Sun, H.; Yang, W.-L.; Deng, W.-P. *Sci. China Chem.* **2020**, *63*, 911–916.

116. See for instance: (a) See for example: Ciochina, R.; Grossman, R. B. *Chem. Rev.* **2006**, *106* (9), 3963–3986; (b) Li, G.; Kusari, S.; Spoteller, M. *Nat. Prod. Rep.* **2014**, *31*, 1175–1201; (c) Yang, X.-W.; Grossman, R. B.; Xu, G. *Chem. Rev.* **2018**, *118* (7), 3508–3558.

117. See for instance: (a) See for instance: Ioffe, S. L., Feuer, H., Eds. *Nitrile Oxides, Nitrones, and Nitronates Inorganic Synthesis: Novel Strategies in Synthesis*; Wiley: Hoboken, 2008; pp. 435–747; (b) Denmark, S. E.; Thorarensen, A. *Chem. Rev.* **1996**, *96*, 137–165; (c) Lubin, H.; Dupuis, C.; Pytkowicz, J.; Brigaud, T. *J. Org. Chem.* **2013**, *78*, 3487–3492; (d) Wolf, C.; Xu, H. *Chem. Commun.* **2011**, *47*, 3339–3350; (e) Hirayama, S.; Wada, N.; Kuroda, N.; Iwai, T.; Yamaotsu, N.; Hirono, S.; Fujii, H.; Nagase, H. *Bioorg. Med. Chem. Lett.* **2014**, *24*, 4895–4898; (f) Sharif, E. U.; O'Doherty, G. A. *Eur. J. Org. Chem.* **2012**, *11*, 2095–2108; (g) Chiba, H.; Oishi, S.; Fujii, N.; Ohno, H. *Angew. Chem. Int. Ed.* **2012**, *51*, 9169–9172; (h) Scott, J. D.; Williams, R. M. *Angew. Chem. Int. Ed.* **2001**, *40*, 1463–1465; (i) Lubin, H.; Tessier, A.; Chaume, G.; Pytkowicz, J.; Brigaud, T. *Org. Lett.* **2010**, *12*, 1496–1499; (j) Agami, C.; Couty, F. *Eur. J. Org. Chem.* **2004**, *2004*, 677–685.

118. Zhao, C.; Shah, B. H.; Khan, I.; Kan, Y.; Zhang, Y. *J. Org. Lett.* **2019**, *21*, 9045–9049.

119. Park, J.-U.; Ahn, H.-I.; Cho, H.-J.; Xuan, Z.; Kim, J.-H. *Adv. Synth. Catal.* **2020**, *362*, 1836–1840.

120. See for instance: (a) See for example: Yang, W.; Zhang, W. *Chem. Soc. Rev.* **2015**, *47*, 1783–1810; (b) Qin, T.; Jiang, Q.; Ji, J.; Luo, J.; Zhao, X. *Org. Biomol. Chem.* **2019**, *17*, 1763–1766; (c) Mollo, M. C.; Orelli, L. R. *Org. Lett.* **2016**, *18*, 6118–6119.

121. See for instance: (a) See for instance: Kaneko, H.; Takahashi, S.; Kogure, N.; Kitajima, M.; Takayama, H. *J. Org. Chem.* **2019**, *84*, 5645–5654; (b) Shi, Y.; Wang, Q.; Gao, S. *Org. Chem. Front.* **2018**, *5*, 1049–1066; (c) Dhanjee, H. H.; Kobayashi, Y.; Buergler, J. F.; McMahon, T. C.; Haley, M. W.; Howell, J. M.; Fujiwara, K.; Wood, J. L. *J. Am. Chem. Soc.* **2017**, *139*, 14901–14904.

122. Jiang, F.; Yuan, F.-R.; Jin, L.-W.; Mei, G.-J.; Shi, F. *ACS Catal.* **2018**, *8*, 10234–10240.

123. Ahn, H.-I.; Park, J.-U.; Xuan, Z.; Kim, J. H. *Org. Biomol. Chem.* **2020**, *18*, 9826–9830.

124. See for instance: (a) See for example: Sharma, S.; Srivastava, V. K.; Kumar, A. *Eur. J. Med. Chem.* **2002**, *37*, 689–697; (b) Matos, M. A. R.; Miranda, M. S.; Morais, V. M. F.; Liebman, J. F. *Eur. J. Org. Chem.* **2004**, *15*, 3340–3345;

(c) Domene, C.; Jenneskens, L. W.; Fowler, P. W. *Tetrahedron Lett.* **2005**, *46*, 4077–4080; (d) Chauhan, J.; Fletcher, S. *Tetrahedron Lett.* **2012**, *53*, 4951–4954; (e) Han, S.-J.; Vogt, F.; May, J. A.; Krishnan, S.; Gatti, M.; Virgil, S. C.; Stoltz, B. M. *J. Org. Chem.* **2015**, *80*, 528–547.

125. Cheng, Q.; Xie, J.-H.; Weng, Y.-C.; You, S.-L. *Angew. Chem. In. Ed.* **2019**, *58*, 5739–5743.

126. See for instance: (a) See for instance: Galliford, C. V.; Scheidt, K. A. *Angew. Chem. Int. Ed.* **2007**, *46*, 8748–8758; (b) Marti, C.; Carreira, E. M. *Eur. J. Org. Chem.* **2003**, *2003*, 2209–2219; (c) Trost, B. M.; Brennan, M. K. *Synthesis* **2009**, *2009*, 3003–3025; (d) Peddibhotla, S. *Curr. Bioact. Compd.* **2009**, *5*, 20–38.

127. Niu, B.; Wu, X.-Y.; Wei, Y.; Shi, M. *Org. Lett.* **2019**, *21*, 4859–4863.

128. See for instance: (a) See for instance: Brown, J. D. In *Comprehensive Heterocyclic Chemistry*; Katrizky, A. R., Rees, C. W., Eds.; Vol. 3; Pergamon Press: Oxford, 1984; p. 57; (b) Evans, P. A.; Holmes, A. B. *Tetrahedron* **1991**, *47*, 9131–9166; (c) Bennasar, M.-L.; Zulaica, E.; Sole, D.; Alonso, S. *Chem. Commun.* **2009**, 3372–3374; (d) Toma, T.; Kita, Y.; Fukuyama, T. *J. Am. Chem. Soc.* **2010**, *132*, 10233–10235; (e) Fukuyama, T.; Xu, L.; Goto, S. *J. Am. Chem. Soc.* **1992**, *114*, 383–385.

129. Gao, C.; Wang, X.; Liu, J.; Li, X. *ACS Catal.* **2021**, *11*, 2684–2690.

130. See for instance: (a) See for example: Molander, G. A. *Chem. Res.* **1998**, *31*, 603–609; (b) Yet, L. *Chem. Rev.* **2000**, *100*, 2963–3007; (c) Deiters, A.; Martin, S. F. *Chem. Rev.* **2004**, *104*, 2199–2238; (d) Hoveyda, A. H.; Zhugralin, A. R. *Nature* **2007**, *450*, 243–251; (e) Hussain, A.; Yousuf, S. K.; Mukherjee, D. *RSC Adv.* **2014**, *4*, 43241–43257; (f) Marti-Centelles, V.; Pandey, M. D.; Burguete, M. I.; Luis, S. V. *Chem. Rev.* **2015**, *115*, 8736–8834.

131. See for instance: (a) Some examples: Still, W. C.; Galynker, I. *Tetrahedron* **1981**, *37*, 3981–3996; (b) Still, W. C.; Novack, V. *J. Am. Chem. Soc.* **1984**, *106*, 1148–1149; (c) Xu, Z.; Johannes, C. W.; Salman, S. S.; Hoveyda, A. H. *J. Am. Chem. Soc.* **1996**, *118*, 10926–10927.

132. Rong, Z.-Q.; Yang, L.-C.; Liu, S.; Yu, Z.; Wang, Y.-N.; Tan, Z. Y.; Huang, R.-Z.; Lan, Y.; Z. Y. *J. Am. Chem. Soc.* **2017**, *139*, 15304–15307.

133. Uno, H.; Punna, N.; Tokunaga, E.; Shiro, M.; Shibata, N. *Angew. Chem. Int. Ed.* **2020**, *59*, 8187–8194.

134. See for instance: (a) See for instance: Majhi, T. P.; Achari, B.; Chattopadhyay, P. *Heterocycles* **2007**, *71*, 1011–1052; (b) Zhu, R.; Wei, J.; Shi, Z. *Chem. Sci.* **2013**, *4*, 3706–3711; (c) Palazzo, T. A.; Mose, R.; Jørgensen, K. A. *Angew. Chem. Int. Ed.* **2017**, *56*, 10033–10038.

135. Wang, Y.-N.; Yang, L.-C.; Rong, Z.-Q.; Liu, T.-L.; Liu, R.; Zhao, Y. *Angew. Chem. Int. Ed.* **2018**, *57*, 1596–1600.

136. Zhang, Q.-L.; Xiong, Q.; Li, M.-M.; Xiong, W.; Shi, B.; Lan, Y.; Lu, L.-Q.; Xiao, W.-J. *Angew. Chem. Int. Ed.* **2020**, *132*, 14200–14204.

About the authors

Pol De La Cruz-Sànchez was born in Lleida, Spain, in 1995. She received his BS in chemistry from Universitat Rovira i Virgili in 2016. She did the final project degree of Chemistry under the supervision of Prof. Anna Maria Masdeu working in the use of CO_2 in metal catalyzed processes using 4-N donor ligands. He completed his Master Degree in "Synthesis, Catalysis and Molecular Design" in 2018 in the Universitat Rovira i Virgili. He did the final Master project in the synthesis, characterization and application of chiral thioeter-carbene ligands, under the supervision of Prof. Montserrat Diéguez. He is currently doing his PhD under the supervision of Prof. Montserrat Diéguez and Prof. Oscar Pàmies. His research is focused on the design, synthesis and screening of modular ligand libraries and their application in several asymmetric metal catalyzed reactions.

Oscar Pàmies obtained his PhD in Prof. Claver's group in 1999 at the Rovira i Virgili University. After three years of post-doctoral work in the group of Prof. J.-E. Bäckvall at Stockholm University, he returned to Tarragona in 2002. He is currently working as professor at the Universitat Rovira i Virgili. He received the Grant for Research Intensification from URV in 2008. He has been awarded the ICREA Academia Prize 2010 from the Catalan Institution for Research and Advanced Studies. His research interests are asymmetric catalysis, water oxidation, enzyme catalysis, organometallic chemistry, and combinatorial synthesis.

CHAPTER FOUR

Evolution in the metal-catalyzed asymmetric hydroformylation of 1,1′-disubstituted alkenes

Jèssica Margalef*, Joris Langlois, Guillem Garcia, Cyril Godard, and Montserrat Diéguez*

Departament de Química Física i Inorgànica, Universitat Rovira i Virgili, C/Marcel·lí Domingo, Tarragona, Spain
*Corresponding authors: e-mail address: montserrat.dieguez@urv.cat; jessica.margalef@urv.cat

Contents

1. Introduction	182
2. Regioselectivity on the asymmetric hydroformylation of 1,1′-disubstituted alkenes	184
3. Asymmetric hydroformylation of 1,1′-disubstituted alkenes with coordinative groups	185
3.1 *N*-selective asymmetric hydroformylation	185
3.2 *i*-Selective asymmetric hydroformylation	195
4. Asymmetric hydroformylation of 1,1′-disubstituted alkenes with non-coordinative groups	199
5. Rationalization of the catalyst efficiency	207
6. Conclusions and outlook	210
Acknowledgments	211
References	211
About the authors	213

Abstract

Despite remarkable progress over the past decade, asymmetric hydroformylation (AHF) is still an underdeveloped field compared to other enantioselective transformations involving alkenes and thus has not encountered a place at an industrial scale yet. The research in AHF has been mainly focused on monosubstituted and 1,2-disubstituted substrates and many excellent reviews about its progress have been published. Less attention has been paid to the AHF of 1,1′-disubstituted alkenes although in the last years significant advances were made in expanding substrate scope and catalyst design. This review focuses on the progress made in the AHF of 1,1′-disubstituted alkenes, from the most studied alkenes bearing coordinating groups to the most challenging unfunctionalized ones.

Advances in Catalysis, Volume 69
ISSN 0360-0564
https://doi.org/10.1016/bs.acat.2021.11.004

Copyright © 2021 Elsevier Inc.
All rights reserved.

1. Introduction

Hydroformylation is an atom-economical way to form aldehydes from readily available feedstocks, namely syngas (a mixture of CO and H_2) and olefins. The versatility of the obtained aldehydes makes this reaction interesting for organic synthesis, since they can be easily converted to other valuable building blocks, such as alcohols, amines, carboxylic acids or nitriles. Hence, oxo-alcohols, which is the name given to the resulting aldehyde products, have a wide range of industrial applications. For example, they can be used for the manufacturing of plasticizers, as chemical intermediates, and for solvent formulations, among others. The hydroformylation of alkenes is therefore one of the most industrially applied transformations catalyzed by homogeneous metal-catalysts [1]. About 10 million tons of oxo-products are produced per year, and this amount is expected to rise nearby 5% during 2021–26 [1,2].

The asymmetric hydroformylation (AHF) reaction thus offers direct access to chiral aldehydes. However, despite the remarkable progress made over the past decade, AHF has not encountered yet a place at an industrial scale and is still underdeveloped compared to other asymmetric transformations involving alkenes [3] such as hydrogenations [4] or hydroborations [5]. The current drawbacks for the practical application of AHF are: interesting catalytic performances were only reported for a reduced range of substrates, there are only a few validations of efficient catalysts immobilization/ recyclability and flow processes, and the reaction mechanism is not well understood yet. There are many difficulties associated with this transformation. Besides the general chemoselectivity issues arising from direct competition between hydroformylation and hydrogenation or isomerization processes, the simultaneous control of regio- and enantioselectivity is often challenging. The fact that regioselectivity is difficult to control in some specific substrates explains its limited substrate scope. The most successful substrates for AHF are mono-substituted alkenes containing an electron withdrawing substituent that induces branched-selectivity by creating two olefinic carbons with different electronic properties. Thus, styrenes and vinyl acetates can be nowadays exclusively hydroformylated into the branched-product exclusively with enantioselectivities higher than 90% ee [3]. Symmetrical alkenes like norbornene and stilbene derivatives are other substrates for which the regioselectivity is also easier to control [3]. In contrast, the AHF of simple alkenes without electron withdrawing groups and 1,2-disubstituted olefins are still a challenge [3].

In the current state of the art, a specific ligand is needed for each type of substrate (Fig. 1).[3d] For example, while Binaphos (**L3**) works best for styrene, Ph-BPE (**L5**) is optimal for allyl cyanide and BOBphos (**L4**) is the ligand of choice for 1-alkyl alkenes. The only ligand with somewhat diverse applicability has been the Bisdiazaphos (**L6**) that works with styrene, vinyl acetate and challenging 1,2-disubstituted olefins.

Other challenging substrates are the 1,1′-disubstituted alkenes. According to the empirical Keuleman's rule *(6,7)*, the main product formed in their AHF is the linear aldehyde, mainly due to increased steric bulk at the α-position of the olefin. However, reaching high enantioselectivities in the AHF of 1,1′-disubstiuted alkenes is tricky because the steric difference between the two substituents of the olefinic carbon makes it difficult to differentiate between each coordinative face of the olefin. This review focuses on the AHF of 1,1′-disubstituted olefins. Since the excellent review about the AHF of this challenging type of substrates was published in 2015 *(3b)*, some remarkable advances were made in the field, mainly with unfunctionalized alkenes as substrates. Two recent short reviews on the efforts in the AHF were published *(3d,e)* with updated examples about 1,1′-disubstituted substrates, although they are only briefly discussed and miss the progress in ligand design and substrate scope that a review covering only 1,1′-disubstituted substrates could provide. We therefore report here an updated overview about the progress made in the field, from the most studied alkenes bearing coordinating

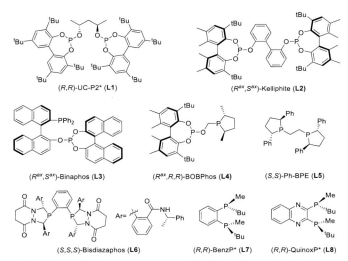

Fig. 1 A selection of the most successful ligands used in the asymmetric hydroformylation.

groups to the most challenging unfunctionalized ones. The review is divided in five sections. After short Section 2 about the regioselectivity issues of AHF of 1,1′-disubstituted substrates, Sections 3 and 4 collect the examples reported about the AHF of 1,1′-disubstituted alkenes considering whether the substrates contain a coordinative group or not. In Section 5, we discuss the key ligand parameters for high selectivities. Initial attempts were performed using mixed Pt/Sn, but more recently, only Rh-systems were utilized. The first ligands tested were chiral diphosphines but later, higher selectivities were achieved with diphospholanes, such as the Ph-BPE (**L5**), and P-stereogenic diphosphines, such as BenzP* (**L7**) or QuinoxP* (**L8**). More recently, it has been found that P-phosphoramidite ligands are interesting candidates for this reaction, providing unprecedented high enantioselectivities for some type of 1,1′-disubtituted olefins.

2. Regioselectivity on the asymmetric hydroformylation of 1,1′-disubstituted alkenes

In 1948, Keuleman stated that "addition of the formyl group to a tertiary carbon atom does not occur, so that no quaternary C atoms are formed" *(6)*. According to this empirical rule, the steric bulk at the α-position of the olefin favors the addition of the formyl group onto the less crowded terminal carbon, and therefore the AHF of 1,1′-disubstituted alkenes would only provide the linear product and avoid any regioselectivity issues (Scheme 1) *(7)*. Most of the reported AHF reactions discussed in this review support this rule. However, examples of racemic hydroformylation of 1,1′-disubstituted alkenes giving exclusively the branched aldehyde have been documented *(8)* and in few cases, the stereo controlled quaternary center could be even obtained (see below in Section 3.2). Indeed, it has been observed that the branched aldehyde can be formed using certain substrates holding highly electron withdrawing substituents and hindered ligands. This finding is highly appealing because the branched product possesses a stereogenic tetrasubstituted center, which has increased the interest of finding catalysts and substrates that can preferentially produce α-tetrasubstituted aldehydes.

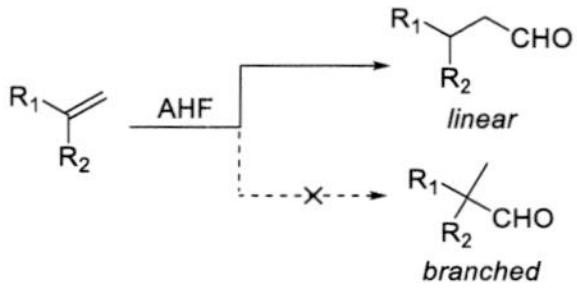

Scheme 1 Regioselectivity achieved in the asymmetric hydroformylation according to Keuleman's rule.

3. Asymmetric hydroformylation of 1,1'-disubsituted alkenes with coordinative groups

In this Section, we discuss the progress made in the AHF of 1,1'-disubstituted alkenes with coordinative groups according on the type of selectivity achieved (*n*-selectivity to the linear aldehyde; see Section 3.1 or *i*-selectivity to the branched aldehyde; see Section 3.2).

3.1 N-selective asymmetric hydroformylation
α-Substituted acrylate and acrylamide derivatives.

The first reported examples of the AHF of 1,1'-disubstituted olefins appeared in 1987 and were focused on 2-substituted acrylates (Scheme 2) *(9,10)*. In both reports, a combined Pt/Sn catalytic system and a chiral diphosphine ligand were employed. Kollár and co-workers were able to obtain a promising enantioselectivity of 82% ee in the AHF of dimethyl itaconate with the PtCl(SnCl$_3$)[(*R,R*)-DIOP] catalytic system *(9)*. Unfortunately, when other unsaturated esters were tested, enantioselectivities were lower (ranging from 40% to 56% ee, Scheme 2A). While the system showed full regioselectivities for hydroformylation at the terminal methylene group, hydrogenation of the substrate could not be avoided, resulting in low yields in most cases (21–94% yield). It should be pointed out that (*R,R*)-DIOP (**L9**) was also tested in combination with [Rh(CO)$_2$Cl$_2$] in the AHF of dimethyl itaconate *(11)*. In contrast to the PtCl$_2$-SnCl$_2$ catalytic system, similar amount of the

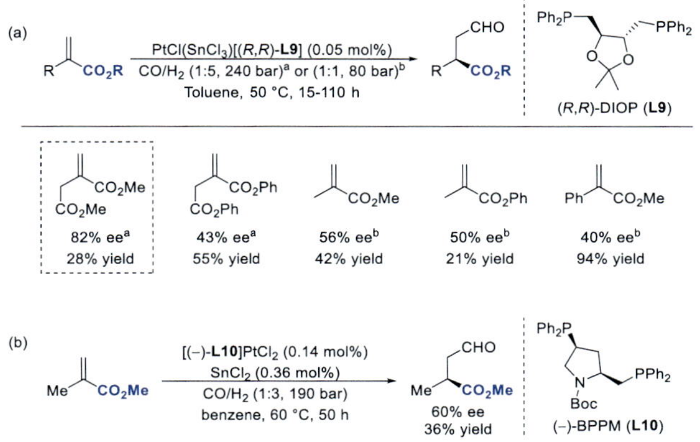

Scheme 2 First Pt/Sn-catalytic systems for the asymmetric hydroformylation of 1,1'-disubstituted acrylates.

linear and branched aldehydes were obtained (43% vs 40%). Furthermore, both products were yielded with very low enantioselectivities (8% and 1% ee for the linear and the branched, respectively). The same year, Stille tested a ((−)-BPPM)PtCl$_2$ chiral catalyst combined with SnCl$_2$ in the AHF of methacrylate, affording the corresponding linear aldehyde with 60% ee (Scheme 2B) (10). Again, a low yield was afforded but in this case it was due to a low branched/linear ratio. It should be noted that the same catalyst provided higher enantioselectivities (70–80% ee) in the AHF of monosubstituted vinylic substrates, such as styrenes, vinyl acetates or N-vinylphthalimide (10), reinforcing that in AHF for each type of substrate, a specific ligand must be developed.

Besides the poor yields, the low substrate scope (being mostly limited to dimethyl itaconate and methacrylate) and modest enantioselectivities obtained, the systems of Kollár and Stille required high hydrogen pressure (80–240 bar) and a very long reaction time (typically 45–110 h). It was not until 2011 that Buchwald et al. reported the first AHF of various α-alkyl acrylates performed under milder reaction conditions using a Rh-catalyst (Scheme 3) (12). Both the pressure and the reaction time could be significantly reduced to 10 bar and 4–8 h, respectively. However, the temperature had to be increased up to 100 °C. By screening a wide range of earlier ligands that were efficient for mono- and 1,2-disubstituted olefins (e.g., Binaphos, diazaphospholane, and Kelliphite, etc.), they found that the use of P-stereogenic diphosphine ligands was crucial to achieve the highest enantioselectivities. Concretely, (R,R)-BenzP* (L7) and (R,R)-QuinoxP* (L8) (Fig. 1) provided 82% and 75% ee in the AHF of ethyl 2-benzylacrylate. With (R,R)-BenzP*, a selection of the desired chiral linear aldehydes were obtained with unprecedented

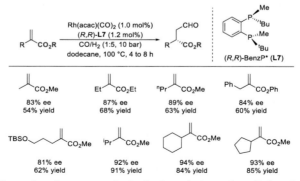

Scheme 3 Rh-catalyzed asymmetric hydroformylation of α-alkylacrylates with the P-chirogenic phosphine ligand (R,R)-BenzP* (L7).

enantioselectivities (81–94% ee) (Scheme 3). In addition, the authors found the optimal CO/H_2 ratio of 1:5 to avoid side reactions and achieve good chemo- and regioselectivities. The highest yields and enantioselectivities were obtained for alkenes bearing secondary alkyl substituents, such as isopropyl, cyclohexyl, and cyclopentyl groups (84–91% yield and 92–94%). It was suggested that one of the reasons for the better yields achieved with these substrates is the presence of bulkier secondary alkyl groups that hampers the formation of the branched Rh-alkyl intermediate, resulting in a better regioselectivity toward the chiral linear aldehydes.

The success of (R,R)-BenzP* (**L7**) and (R,R)-QuinoxP* (**L8**) in the Rh-catalyzed AHF of acrylates *(12)*, prompted Godard and co-workers to test them in the asymmetric intermolecular hydroaminomethylation (HAM) of α-alkyl acrylates *(13)*. In this case, (R,R)-QuinoxP* (**L8**) showed the best catalytic performance. For the first time, a single catalyst allowed the efficient and straightforward synthesis of chiral γ-aminobutyric esters from readily available acrylates. Different acrylates and amines were subjected to the HAM reaction releasing a range of synthetically valuable chiral γ-aminobutyric esters with ee's up to 86% (Scheme 4). As for the Buchwald's system, the best results were achieved with substrates containing bulky secondary alkyl groups.

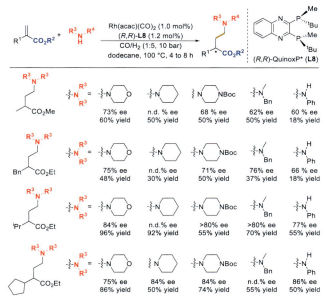

Scheme 4 Rh-catalyzed asymmetric intermolecular hydroaminomethylation of α-alkyl acrylates with (R,R)-QuinoxP* (**L8**).

Much after the Buchwald achievement, in 2019, Chin et al. reported the application of pyrrolylphosphinite ligand **L11**, without a stereogenic P group, in the Rh-catalyzed AHF of α-substituted acrylates *(14)*. A range of chiral aldehydes were achieved in good-to-high enantioselectivities (73–86% ee) (Scheme 5). Although, the enantioselectivities were somewhat lower than those reported with the Rh-(R,R)-BenzP* system, the system also operated at low pressure (10 bar) but at a lower temperature (80 °C), with high productivity (turnover number (TON) up to 8900), excellent regioselectivity, and good-to-high chemoselectivities, which resulted in good yields (up to 87%).

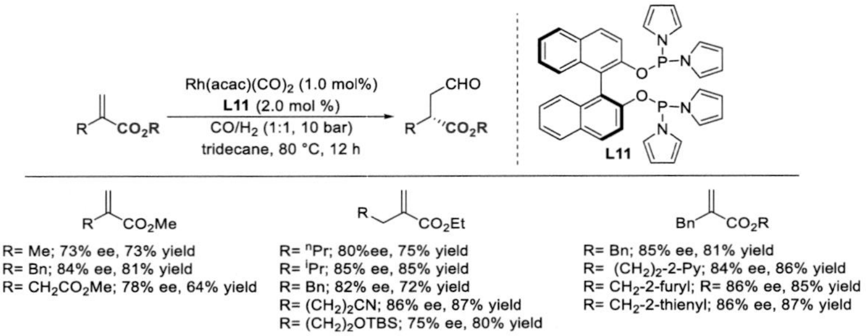

Scheme 5 Rh-catalyzed asymmetric hydroformylation of α-alkylacrylates with pyrrolylphosphinite ligand **L11**.

The most recent contribution in the AHF of α-substituted acrylates has been made by Zhang and co-workers with a phosphine-phosphoroamide ligand from the (S,S)-YanPhos family *(15)*. This ligand family was previously successfully applied on the AHF of 1,1′-disubstituted alkenes with weakly coordinative groups *(16)* and without coordinative groups *(17)* providing high activities and selectivities (see in Section 4). They initially studied the AHF of ethyl 2-benzylacrylate by testing two (S,S)-Yanphos ligands, with distinct substituents in the phosphine moiety (Ar = 3,5-^{t}Bu-4-MeO-C_6H_3 and 3,5-^{t}Bu-C_6H_4, Scheme 6), and a range of representative ligands that were efficient for AHF and hydrogenation, such as (S^{ax},R)-Duanphos, XuPhos, (S,S)-Me-DuPhos and Walphos. They found that the (S,S)-DTBM-YanPhos ligand (**L12**) provided the best catalytic performance. Advantageously, they could even reduce the H_2 pressure down to only 5 bars, maintaining the temperature used in the previous example. With this ligand, a range of linear aldehydes with β-chirality (Scheme 6) could be obtained with high enantioselectivities (88–96% ee) and high yields (up to 94%). In contrast to (R,R)-BenzP* (**L7**) and (R,R)-QuinoxP* (**L8**), a slightly lower enantioselectivity was observed with a substrate bearing a bulkier isopropyl group

Metal-catalyzed asymmetric hydroformylation of 1,1′-disubstituted alkenes 189

on the α-position (73% ee). The authors also showed that the AHF of α-methyl methacrylate could be performed at a Gram-scale and at a lower catalyst loading of 0.05 mol%, without compromising the yield nor enantioselectivity. The utility of the methodology was further demonstrated by the concise synthetic route to chiral γ-butyrolactone (Scheme 7).

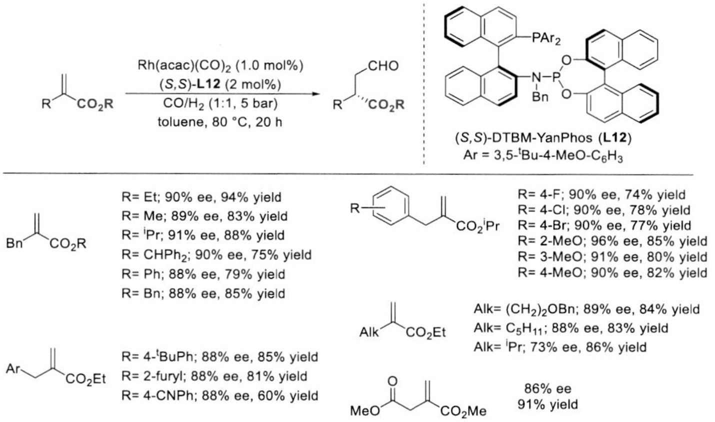

Scheme 6 Rh-catalyzed asymmetric hydroformylation of α-substituted acrylates with (S,S)-DTBM-YanPhos (**L12**).

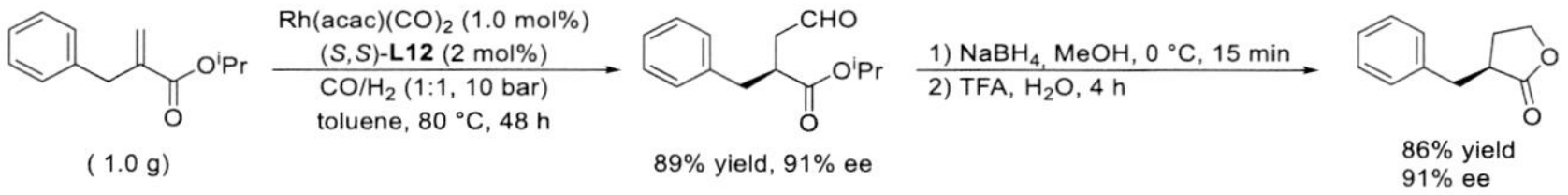

Scheme 7 Synthesis of chiral γ-butyrolactone at a 1.0 g scale with (S,S)-DTBM-YanPhos (**L12**).

Despite the similarity of α-substituted acrylamides with α-acrylates, the AHF of the former was not explored until very recently. Indeed, only a few examples were reported in the asymmetric hydroformylation of non-substituted acrylamides with moderate yields and enantioselectivities *(18)*. The only two existing publications reporting the AHF of α-substituted acrylamides appeared in 2020 *(15,19)*. In one of them, only three substrates were studied. In particular, three 2-benzylacrylamides, with different amide substituents were tested with the Rh/(S,S)-DTBM-YanPhos (**L12**) complex (Scheme 8) *(15)*. While acrylamides with cyclic amide substituents provided good levels of enantioselectivity (up to 86% ee), the presence of a diethyl group resulted in a poor enantioselectivity (28% ee). It is interesting to note that a longer reaction time was required compared to the AHF of acrylates with the same catalyst (48 h vs 20 h, Schemes 6 vs 8, respectively).

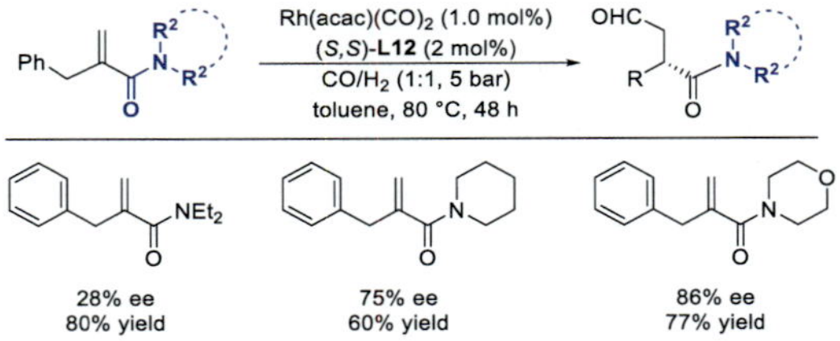

Scheme 8 Asymmetric hydroformylation of acrylamides catalyzed by the Rh/(S,S)-DTBM-YanPhos (**L12**) complex.

The second study about the AHF of α-substituted acrylamides employed 1,3-phosphite-phosphoramidite ligands based on a sugar backbone derived from D-xylose (**L13**, Scheme 9). Their modularity was key to achieve the best catalytic performance, being ligands with a bulky N-group and an (S^{ax}, S^{ax})-biaryl moiety the ones that provided the best results (R=Cy and R=(S)-CHMePh) (Scheme 9). With the optimal ligands, it was possible to hydroformylate for the first time a broad range of acrylamides with different substituents at the α-position and on the amide group with high-to- excellent enantioselectivities (74–99% ee) and good-to-high yields (50–87%) *(19)*. Furthermore, hydroformylation reactions could be performed at 60 °C. For most of the substrates the best results were achieved with Rh/(S^{ax}, S^{ax})-**L13** (R=Cy) catalytic system. For more sterically hindered alkenes (**3–5**) ligand with a (S)-σ-methylbenzyl amine substituent provided the best results. The results for substrates **5** and **9** are relevant since important biologically active molecules include a phenyl substituent in α-position and the catalysts reported to date *(12, 13, 15)* were only efficient for α-alkyl substituted substrates.

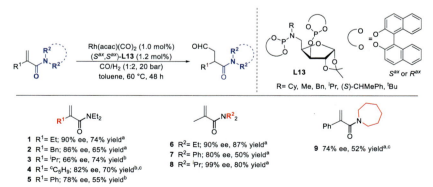

Scheme 9 Rh-catalyzed asymmetric hydroformylation of α-substituted acrylamides with sugar-based phosphite-phosphoramidite ligands **L7**. [a]Ligand (S^{ax}, S^{ax})-**L13** (R=Cy) was used. [b]Ligand (S^{ax}, S^{ax})-**L13** (R=(S)-CHMePh) was used. [c]Temperature=90 °C for 16 h.

Moreover, $Rh/(S^{ax},S^{ax})$-**L13** ($R = Cy$) successfully promoted the one-pot asymmetric HAM of different α-substituted acrylamides and amines to directly yield chiral γ-aminobutyric acid (GABA) derivatives, with also high yields and enantioselectivities (Scheme 10). The practicality of the reaction was shown by synthesizing the brain imaging molecule RWAY in one single step with high enantioselectivity (Scheme 11).

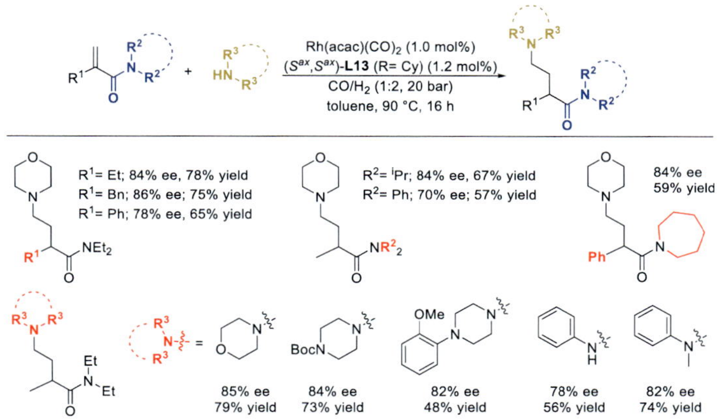

Scheme 10 Rh-catalyzed asymmetric hydroaminomethylation of α-substituted acrylamides with ligand (S^{ax},S^{ax})-**L13** ($R = Cy$).

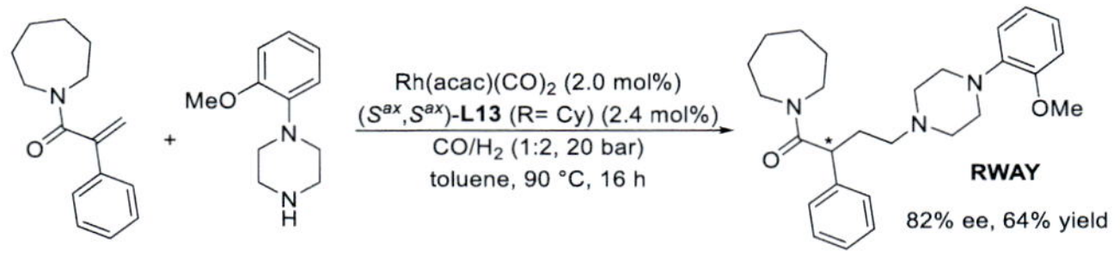

Scheme 11 Synthesis of RWAY via HAM Reaction catalayzed by $Rh/(S^{ax},S^{ax})$-**L13** ($R = Cy$).

3.1.1 Other functionalized 1,1′-disubstituted alkenes

Besides α-substituted acrylates and acrylamides, a few substrates containing a coordinative group other than an ester or an amide at the olefinic carbon have been studied. In 2010, Landis and co-workers explored the AHF of 1,1′-disubstituted ene-phthalimides with a Rh-complex of (S,S,S)-Bisdiazaphos (**L6**) (Scheme 12) *(20)*. The reaction was selective toward the β-aldehyde although variable amounts of the isomerized internal alkene were also observed. Promising results were achieved for the allyl benzyl ether derived ene-phthalimide, which provided a chemoselectivity up to

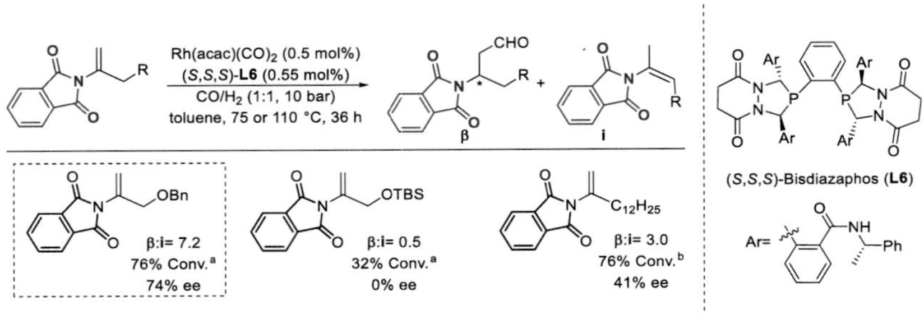

Scheme 12 Rh-catalyzed asymmetric hydroformylation of 1,1′-disubstituted ene-phthalimides with the (S,S,S)-Bisdiazaphos ligand (**L6**). [a]75 °C. [b]110 °C.

7:1 (β-aldehyde/internal alkene) and an enantioselectivity of 74% ee. Unfortunately, poorer chemo- and enantioselectivities were observed when other derivatives were used. Nevertheless, this reaction constitutes a straightforward way to prepare chiral β^3-aminoaldehydes.

Börner et al. were the first to explore the AHF of α-phosphorylated vinyl arenes as an alternative route toward 3-aryl-3-phosphorylated propanals *(21)*. An extensive ligand screening (e.g., (S,S)-DIOP (**L9**), (R,R)-QuinoxP* (**L7**), (S,S)-BenzP* (**L8**), (R,R,S)-Bisdiazaphos (**L6**), (R,R)-Kelliphite (**L2**)) revealed sugar-based ligands **L13** (Scheme 9) as the best candidates for this transformation. In particular, **L13** (R = Bn) containing an (R^{ax})-binaphtyl group on the phosphite and phosphoramidite moieties provided the highest yields and enantioselectivities. The β-chiral aldehydes were obtained as the major products and only small amounts of hydrogenated products were observed, so high yields were in general achieved (up to 99%) (Scheme 13). Unfortunately, the enantioselectivity was only moderate and not determined

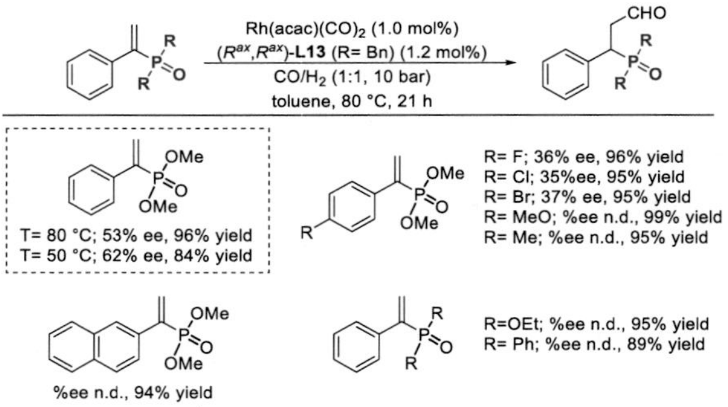

Scheme 13 Rh-catalyzed asymmetric hydroformylation of α-phosphorylated vinyl arenes with ligand (R^{ax}, R^{ax})-**L13** (R = Bn).

for all the substrates tested due to issues in finding a suitable method for their separation. The maximum enantioselectivity obtained was 53% ee, which could be raised up to 62% by decreasing the temperature to 50 °C, at the expanse of the yield (96% vs 84%).

Another group of functionalized 1,1′-terminal alkenes are substituted allylic compounds. The first substituted allylic substrate tested was 3-methylbut-3-enoate with the Pt/Sn/(*R*,*R*)-DIOP system developed by Kollár and co-workers *(9)*. The chiral linear aldehyde was afforded in 85% yield, however, a poor enantioselectivity was obtained (10% ee) (Scheme 14A). Note that among all 1,1′-substituted alkenes tested with Kollár's system (see Scheme 2A), the substituted allylic substrate shown in Scheme 14A is the substrate that gave the lowest ee value.

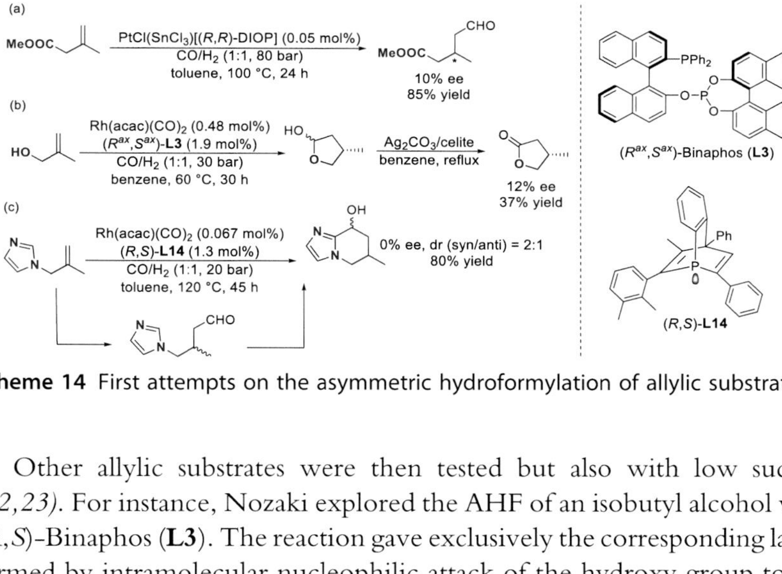

Scheme 14 First attempts on the asymmetric hydroformylation of allylic substrates.

Other allylic substrates were then tested but also with low success *(22,23)*. For instance, Nozaki explored the AHF of an isobutyl alcohol with (*R*,*S*)-Binaphos (**L3**). The reaction gave exclusively the corresponding lactol formed by intramolecular nucleophilic attack of the hydroxy group to the aldehyde moiety. This product was sequentially subjected to oxidation with Ag_2CO_3 on Celite to afford the desired-butyrolactone (Scheme 14B) *(22)*. In this case, not only the enantioselectivity was low (12% ee), but also a disappointing yield was recorded (37%). Although it was expected that the presence of the hydroxyl group would be beneficial for the stereoinduction due to coordination to Rh, the results obtained with **L3** suggest that maybe this coordination is too weak.

Müller also explored the hydroformylation of N-(β-methallyl)imidazole *(23)*. The authors found that σ-donor ligands were necessary to achieve full conversions, since π-acceptor ligands failed to provide the desired product, most likely due to the coordination of the imidazole group of the substrate to the metal center in view of the poor σ-donating character of the ligands. Thus, the more σ-donating phosphabarrelene ligand (R,R)-**L14** (Scheme 14C) provided the hydroformylated product in almost full conversion at 80 °C. Increasing the temperature up to 120 °C promoted the intramolecular cyclization of formed linear aldehyde to give 8-hydroxy-6-methyl-5,6,7,8-tetrahydroimidazo[1,2-a]pyridine (Scheme 14C). However, the corresponding product was obtained as a racemate and with a low diastereoselectivity of 2:1 (syn/anti).

Better results were later achieved in the AHF of allyl phthalimides. The use of a Rh/(S,S)-Ph-BPE (**L5**) catalyst allowed the synthesis of a series of β^3-aminoaldehydes with enantioselectivities up to 95% ee (Scheme 15) *(24)*. However, the system was sensitive to the length of the substrate's alkyl chain, and such levels of asymmetric induction were only observed with substrates bearing a methyl, ethyl, isopropyl or a cyclohexyl group (Scheme 15). In contrast, moderate enantioselectivities were attained when other substrates were tested (55–77% ee). Similarly, the conversion and the chemoselectivity toward the hydroformylation reaction vs hydrogenation was only high in a few cases. Nevertheless, the authors showed the utility of this transformation to obtain chiral β^3-amino acids and alcohols through oxidation or reduction of the N-phthalimide-protected aldehydes. As an example, 3-(aminomethyl)-4-methylpentanoic acid and 3-(aminomethyl)-4-methylpentan-1-ol were prepared without affecting the stereochemistry.

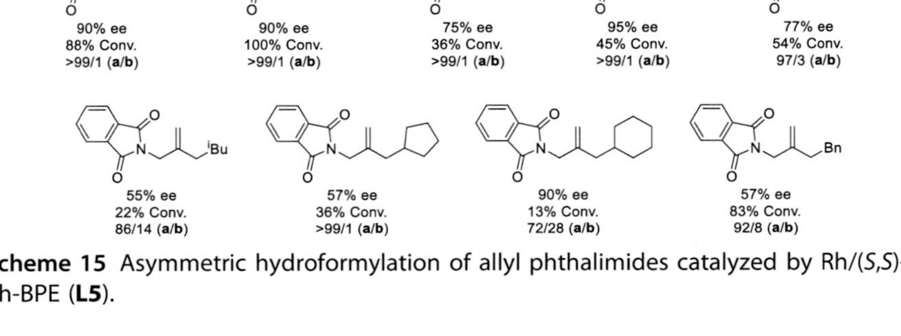

Scheme 15 Asymmetric hydroformylation of allyl phthalimides catalyzed by Rh/(S,S)-Ph-BPE (**L5**).

In contrast with the low enantioselectivity reported by Nozaki in the AHF of an allylic alcohol (Scheme 14B), Zhang's group recently published the successful AHF of allylic alcohols with the Rh/(S,S)-DTBM-YanPhos (**L12**, Scheme 6) system *(16)*. They were able to hydroformylate a broad range of allylic alcohols with enantioselectivities ranging from 85% to 93% ee, independently of the nature of the aryl substituent. The afforded chiral linear aldehydes were subsequently oxidized to the corresponding lactones. It should be pointed out that only aryl-substituted alkenes were tested, which makes difficult to compare the performance of Rh/**L12** with the system used in the previously commented Nozaki's report (Scheme 14B) *(22)*. The AHF of allylic amines revealed more challenging since the corresponding lactams were furnished in lower yields and enantioselectivities (up to 69% yield and 80–86% ee) (Scheme 16). Moreover, the scope of allylic amines was narrow in comparison with the AHF of allylic alcohols.

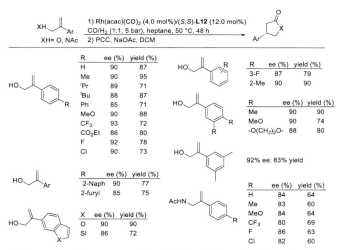

Scheme 16 Rh-catalyzed asymmetric hydroformylation of allylic alcohols and amines with (S,S)-DTBM-YanPhos (**L12**).

3.2 *i*-Selective asymmetric hydroformylation

As previously commented, according to Keulemans' rule, the formation of the linear aldehyde is favored over the branched (see Section 2) *(6,7)*. Nevertheless, in some cases, an unusual regioselectivity was observed and the chiral branched aldehyde was obtained as the major product. Three early examples of *i*-selectivity on the AHF of 1,1′-substituted alkenes were reported in 1995 *(25,26)*. In all of them, the recorded enantioselectivity was insufficient (10–59% ee), but they showed the possibility of preparing chiral aldehydes with a quaternary stereocenter. The most successful example used HRh(CO)(PPh₃)₃ in the presence of a (R,R)-DIOP (**L9**) in the AHF of methyl N-acetamidoacrylate *(25)*. The chiral formyl derivative with a quaternary stereogenic carbon was obtained in 90% yield and with 59% ee (Scheme 17). Although the reaction could be performed at a low temperature (30 °C), a high pressure of 100 bar (CO/H₂ (1:1)) was needed. The authors speculated that polydentate binding of methyl N-acetamidoacrylate to rhodium is the responsible for the regio- and stereo-selectivity observed. More recently, Börner also observed the formation of the branched aldehyde when testing the structurally similar methyl 2-(acetamidomethyl)-acrylate, also using a Rh/DIOP catalyst (Scheme 17B) *(21)*.

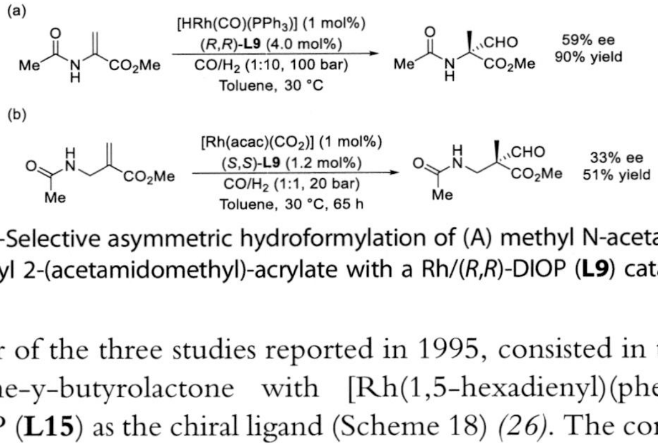

Scheme 17 *i*-Selective asymmetric hydroformylation of (A) methyl N-acetamidoacrylate and (B) methyl 2-(acetamidomethyl)-acrylate with a Rh/(R,R)-DIOP (**L9**) catalyst.

Another of the three studies reported in 1995, consisted in the AHF of α-methylene-γ-butyrolactone with [Rh(1,5-hexadienyl)(phen)]Cl and (R)-BINAP (**L15**) as the chiral ligand (Scheme 18) *(26)*. The corresponding aldehyde lactone with a quaternary chiral center was afforded with an enantioselectivity of only 36% ee that could be increased to 37% ee by decreasing the temperature to 80 °C, but the yield was compromised (15%) (Scheme 18). It should be pointed out that branched lactone was the only product obtained, and any traces of the hydrogenation or isomerization by-products were observed.

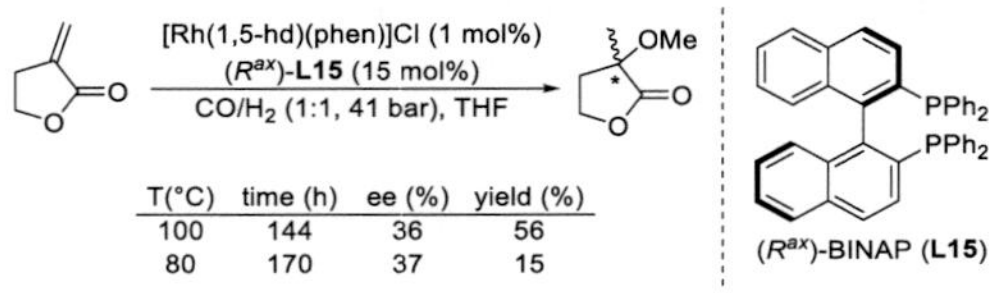

Scheme 18 Rh-catalyzed asymmetric hydroformylation of α-methylene-γ-butyrolactone with (R)-BINAP (**L15**).

The last of the three early works showing *i*-selectivity was found in the AHF of the 1-phenyl-1-(2-pyridyl)ethene (Scheme 16) *(25)*. In this case, the maximum enantioselectivity achieved was ca. 10% ee with a Pydiphos P-oxide ligand (**L16**) (Scheme 19) *(25)*. The AHF of this type of substrates is very challenging because of the little steric difference between both aryl substituents that makes difficult the discrimination between the two prochiral faces. Although low enantioselectivity was achieved, the observed asymmetric induction was postulated to be due to the presence of the *ortho*-pyridyl ring, which could act as a directing group.

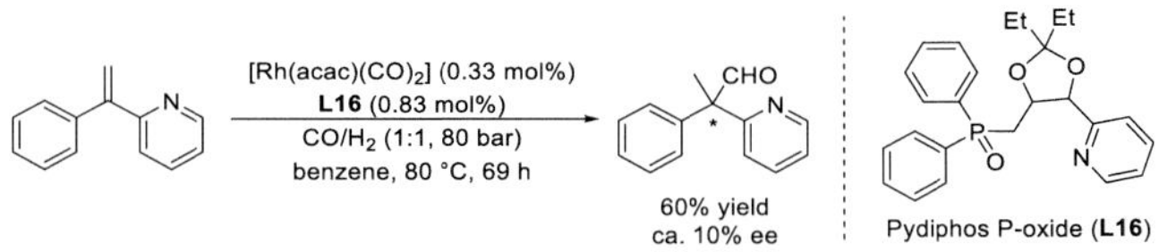

Scheme 19 Rh-catalyzed asymmetric hydroformylation of 1-phenyl-1-(2-pyridyl)ethene with Pydiphos P-oxide ligand (**L16**).

The first successful example providing the branched aldehyde as the major product appears much later, in 2013, with the Buchwald's AHF of 3,3,3-trifluoroprop-1-en-2-yl acetate *(27)*. As in their work about the AHF of α-alkylacrylates, it was found that P-stereogenic ligands provided the best enantioselectivities. Thus, with the (*R*,*R*)-QuinoxP* (**L8**) and (*R*,*R*,*S*,*S*)-Duanphos (**L17**) ligands, the chiral branched aldehyde was obtained with an enantioselectivity as high as 91% and 92% ee, respectively (Scheme 20A). To understand the origin of this unusual regioselectivity, the authors performed the AHF of 2-propenyl acetate under identical reaction conditions. The reaction yielded the chiral linear aldehyde as the major product, suggesting that the presence of the strongly electron-withdrawing trifluoromethyl group favors the branched intermediate. Importantly, the unexpected regioselectivity observed in the AHF of 3,3,3-trifluoroprop-1-en-2-yl acetate allowed the one-pot preparation of the 2-trifluoromethyllactic acid (TFMLA) in excellent enantioselectivity, an important building block bearing a quaternary stereocenter found on many active pharmaceutical compounds (Scheme 20A). Later, Zhang's group reported the AHF of the same substrate, using another P-stereogenic diphosphine ligand, the (*R*)-BIBOP **L18** *(28)*, however with a lower enantioselectivity (80% ee) than those obtained with QuinoxP* or DuanPhos (Scheme 20B).

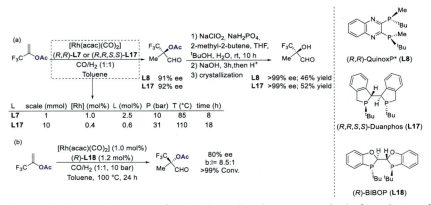

Scheme 20 Reported examples for the Rh-catalyzed asymmetric hydroformylation of 3,3,3-trifluoroprop-1-en-2-yl acetate.

The last example was reported by Landis, Schomaker and co-workers in 2018 *(29)*. By using (S,S,S)-BisDiazaphos (**L6**) or (S,S)-Ph-BPE (**L5**) ligands they could expand the scope of 1,1′-disubstituted alkenes that can be hydroformylated with high branched selectivity under rather mild conditions (10 bar, 60 °C and 2–72 h). In particular, high *i*-selectivities were achieved with some acrylates and enol ester 1,1′-disubstituted alkenes, which contain electron-withdrawing groups, such as an acetate, a fluorine and a trifluoromethyl (Scheme 21). Interestingly, some acrylates lacking electron

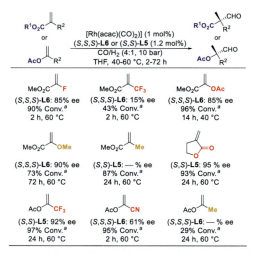

Scheme 21 Selected results for the Rh-catalyzed asymmetric hydroformylation of acrylates and enol ester 1,1′-disubstituted alkenes with (S,S,S)-BisDiazaphos (**L6**) or (S,S)-Ph-BPE (**L5**). [a]Conversion toward the branched aldehyde.

withdrawing groups also provided high *i*-selectivities (e.g., with a OMe or Me). The AHF of some substrates proceeded with high enantioselectivities (up to 95% ee) and with a ratio > 50:1 in favor of the branched product. However, in some cases, the chemoselectivity was low due to the formation of hydrogenation products. Similarly, the enantioselectivity was also dependent on the substrate substituents (Scheme 21).

4. Asymmetric hydroformylation of 1,1′-disubsituted alkenes with non-coordinative groups

As described in the first part, some progress was recently made in the AHF of 1,1′-disubstituted alkenes containing functional groups that might coordinate to the metal center, contributing to achieve the desired high levels of asymmetric induction. Reaching the same value of enantioselectivity is even more difficult when using 1,1′-disubstituted alkenes without a coordinating group. It's only very recently that high enantioselectivities were achieved with certain unfunctionalized substrates.

4.1.1 1,1′-Dialkyl alkenes

The first work on the AHF of purely unfunctionalized 1,1′-dialkyl substituted alkenes was carried out by Gladiali and Botteghi in 1983. Using a Rh catalyst with the DIOP cholesterol derivative ligand (**L19**), they reported the hydroformylation of 2,3,3-trimethyl-butene giving only the linear aldehyde but as a racemate (1% ee, Scheme 22) *(30)*.

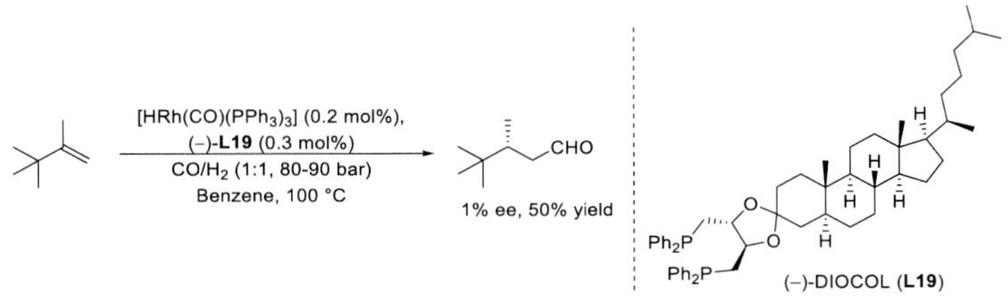

Scheme 22 First attempt of Rh-catalyzed asymmetric hydroformylation of an unfunctionalized 1,1′-dialkyl substituted alkene with (−)-DIOCOL (**L19**).

A few years later, Consiglio and co-workers developed the AHF of 2,3-dimethyl-1-butene with a PtCl$_2$-SnCl$_2$ catalyst and phosphindole chiral ligands. They reached 46% ee toward the (*S*)-product using (*R*,*R*)-BCO-DBP (**L20**) and 36% ee for the opposite enantiomer with the (*R*,*R*)-DIOP-DBP one (**L21**). In both cases a complete control of the regioselectivity was observed (Scheme 23) *(31)*. Despite that the enantioselectivity was improved with respect to the Rh/DIOCOL system, enantioselectivities remained low. Furthermore, harsh reaction conditions of pressure and temperature were required.

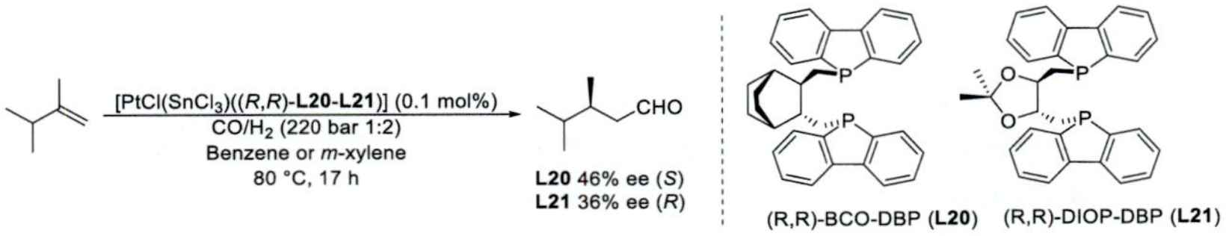

Scheme 23 Pt/Sn-catalyzed asymmetric hydroformylation of 1,1′-dialkyl substituted alkenes with (*R,R*)-BCO-DBP (**L20**) and (*R,R*)-DIOP-DBP (**L21**).

During the investigation of Landis and Schomaker on the *i*-selective AHF of electron deficient acrylates and acetates, they also explored unfunctionalized cycloalkanes as substrates. With (*S,S*)-Ph-BPE (**L5**) it was possible to induce the formation of the branched aldehyde when constrained aryl-substituted cyclopropenes were tested (Scheme 24). The desired chiral branched products were obtained in more than 83% yield and with high diastereoselectivities (up to >19:1 dr). It is interesting to note that the non-asymmetric hydroformylation of electronically unactivated alkenes 7-methylenebicyclo-[4.1.0]heptane and 8-methylenebicyclo-[5.1.0]octane also proceeded with high regioselectivities toward the *i*-aldehyde, in spite

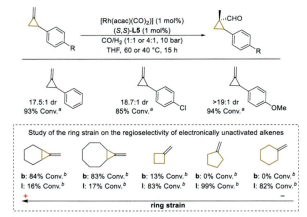

Scheme 24 Rh-catalyzed asymmetric hydroformylation of aryl-substituted cyclopropenes with (*S,S*)-Ph-BPE (**L5**). [a]Conversion toward the branched aldehyde. [b]The product obtained is not chiral.

of not containing any aryl group that could be considered as an electronically activating group. In contrast, the linear product was yielded from electronically unactivated cycloalkanes with ring sizes greater than three, which are therefore less strained rings (Scheme 24). This suggests that ring strain plays an important role in influencing the regioselectivity in the HF of 1,1′-disubstituted alkenes.

Very recently, Zhang's group developed an improved AHF of those very challenging substrates via the introduction of a steric disulfonyl auxiliary, using a Rh/(S,S)-DTB-YanPhos (**L22**) complex as catalyst. The authors proved that the incorporation of the disulfonyl moiety increases the yield of the reaction and drastically enhances the enantioselectivity (Scheme 25A vs B). In addition, a pressure of only 5 bar could be used. This methodology was applied over 20 substrates providing up to 97% yield and > 99% ee (Scheme 25C).

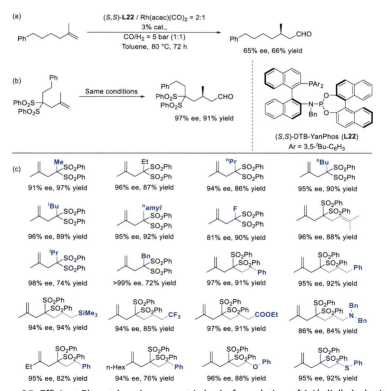

Scheme 25 Efficient Rh-catalyzed asymmetric hydroformylation of 1,1′-dialkylsubstituted alkenes via steric auxiliary help with (S,S)-DTB-YanPhos (**L22**).

Furthermore, the sulfonyl groups can be easily introduced and removed to convert the chiral aldehyde into chiral building blocks with 22–25% overall yield starting from 1,3-benzodithiole tetraoxide, which enhances the synthetic applicability of this transformation *(32)* (Scheme 26). Moreover, the authors showed that the aldehyde group on the chiral products could be readily transformed into other valuable functionalities in a sequential manner followed by the removal of the disulfonyl groups.

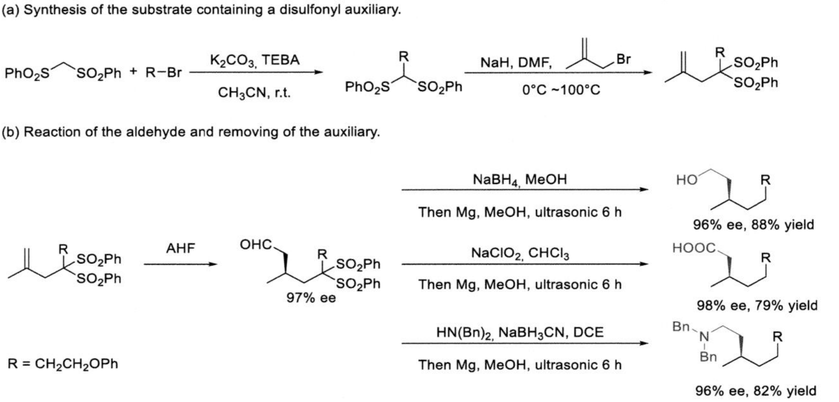

Scheme 26 Synthesis of the substrates containing a disulfonyl auxiliary followed by sequential transformation of the aldehyde.

4.1.2 1-Aryl-1-alkyl alkenes

1-Aryl-1-alkyl substituted alkenes have been more studied than the 1,1′-dialkyl substituted ones, especially α-methylstyrene. Early studies were performed by Consiglio's and Kollár's group using mixed Pt/Sn-catalysts for the AHF of methyl styrene *(33–35)*. Consiglio first tested (S,S)-Chiraphos (**L23**) and (R,R)-DIOP (**L1**), obtaining a maximum of 7% ee with the (S,S)-DIOP ligand (**L9**) (Scheme 27A) *(33)*. Later, the same authors reported the use of (R,R)-DIOP (**L9**) ligand in the AHF of several *para*-substituted styrene derivatives. In this case, the reaction was performed at a higher pressure (180 bar, 1:1 CO/H_2) *(34)*. With this conditions the AHF of methyl styrene proceeded with a slightly better enantioselectivity (15% ee, Scheme 27B). Kollár's group published the AHF of methylstyrene with the (S,S)-BDPP ligand (**L24**) that gave the opposite enantiomer than the obtained with Consiglio's Chiraphos system but under milder conditions, albeit with an even poorer ee value of 9% (Scheme 27C) *(35)*. Moreover a lower yield was obtained due to the low reactivity at 50 °C, since only

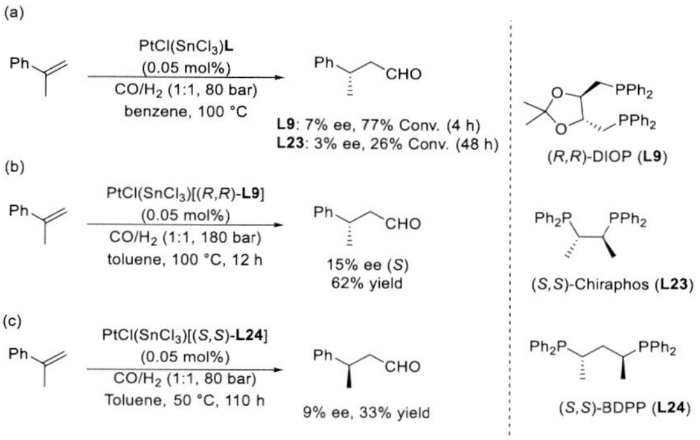

Metal-catalyzed asymmetric hydroformylation of 1,1′-disubstituted alkenes   203

Scheme 27 Pt/Sn-catalyzed asymmetric hydroformylation of α-methylstyrene.

a conversion of 35% was achieved after 110 h. An increase of the temperature resulted in the complete loss of enantioselectivity (ca. 2% ee at 100 °C).

The first Rh-catalyzed AHF of α-methylstyrene was reported by Consiglio's group with the [Rh(*NBD*)Cl]$_2$/(*S,S*)-Chiraphos (**L23**) catalyst (Scheme 28) *(33)*. Compared with the Pt/Sn systems, a higher ee of 21% ee was reached. This low value was the maximum level of enantioselectivity achieved at that time.

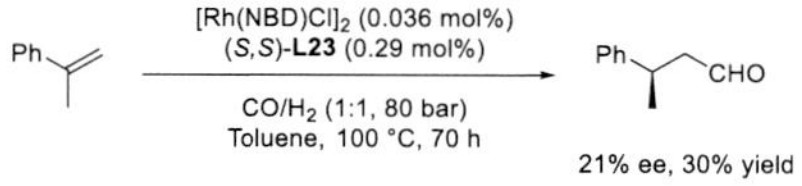

Scheme 28 Rh-catalyzed asymmetric hydroformylation of α-methylstyrene with (*S,S*)-Chiraphos (**L23**).

In 2004, the Rh-catalyzed AHF of α-methylstyrene using the diphosphonite ligand **L25** was patented by the State University of New York and Mitsubishi Chemical Corporation *(36)*. They reported 46% ee but a very low conversion of 15%, which is not surprising considering the low pressure and temperature used (Scheme 29).

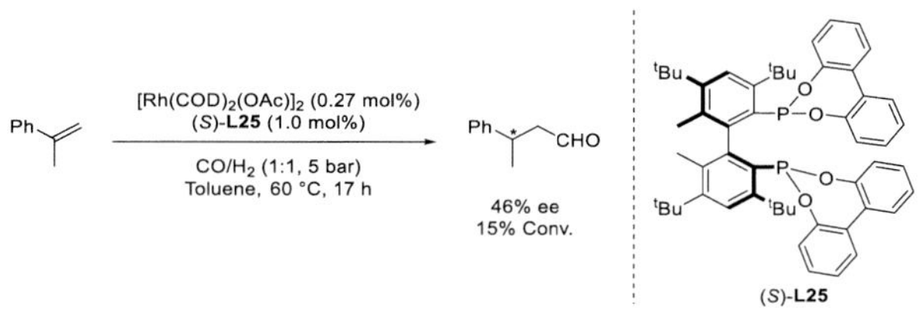

Scheme 29 Rh-catalyzed asymmetric hydroformylation of α-methylstyrene with (*S*)-**L25** as chiral ligand.

Bayón and Pereira applied bulky chiral monophosphite ligands (R,R,R)-**L26** with large cone angle (240–270°) in the AHF of 1,1′-methylstyrene, but unfortunately they obtained an even more disappointing enantioselectivity than the achieved at this point with Rh–catalysts (15% ee, Scheme 30) *(37)*.

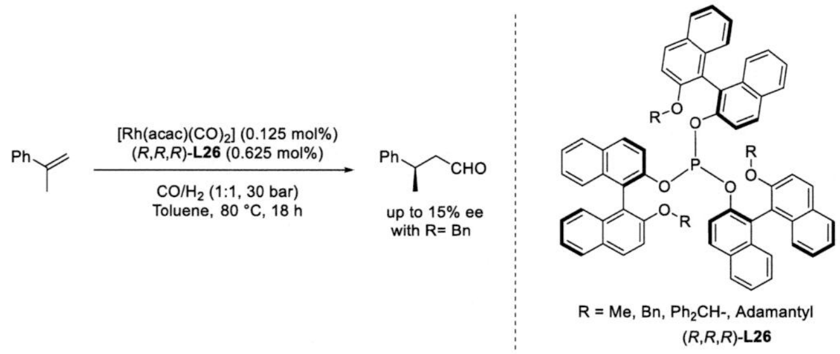

Scheme 30 Rh-catalyzed asymmetric hydroformylation of α-methylstyrene with chiral monophosphite (R,R,R)-**L26**.

Börner et al. tested the family of sugar based phosphite–phosphoramidites ligands **L13** (Scheme 9) in the AHF of α-methylstyrene and reached a moderate enantioselectivity (39% ee) with high yield (95%) (Scheme 31) *(21)*. They observed that the substituent at the N-moiety of the ligand has a crucial effect on the enantioselectivity. Indeed, they observed that sterically demanding substituents such as Cy or ^{t}Bu had a negative effect on the enantioselectivity, which contrasts with the results observed in the AHF of acrylamides (see Scheme 9). Interestingly the best ligand contained a (S)-σ-methylbenzyl moiety at the N-atom and (S)-configuration of the biaryl moieties (see Section 5).

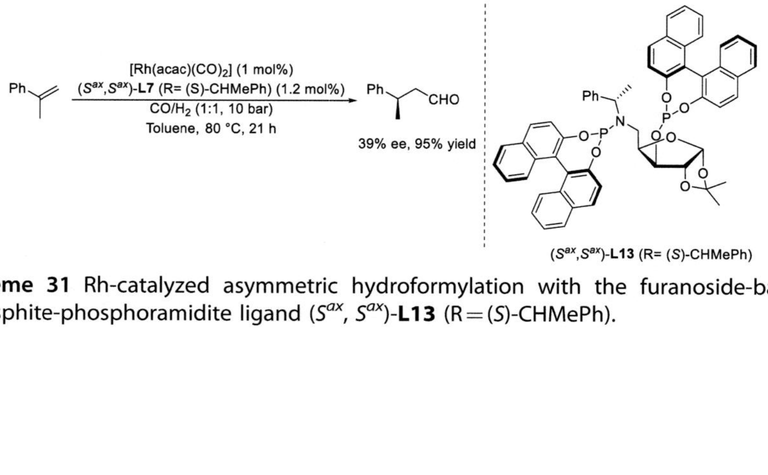

Scheme 31 Rh-catalyzed asymmetric hydroformylation with the furanoside-based phosphite-phosphoramidite ligand (S^{ax}, S^{ax})-**L13** (R = (S)-CHMePh).

In 2018, Zhang group reported the best results so far in the AHF of 1,1′-methylstyrene using Rh(acac)(CO)$_2$ as precatalyst and a YanPhos-type ligand. They observed that lowering the pressure from 20 to 5 bar increases the conversion, thus affording very mild conditions for this reaction. The authors tested a series of YanPhos ligands and reported that the key parameter to achieve high conversion and enantioselectivity is the use of an (S,S)-configured-binol group on the phosphite moiety together with a hindered aryl group on the phosphine. Indeed, when using (S,R)-ligands, a conversion of 5% with 37% ee was obtained, which is very low compared to the results achieved in the preliminary results (65% and 82% ee). After optimization of the ligand and reaction conditions, the AHF of α-methylstyrene was achieved with 87% ee with (S,S)-DTB-YanPhos (**L22**, Scheme 25). With this ligand, a range of 1,1′-alkylstyrene derivatives could be hydroformylated with unprecedented enantioselectivities (up to 92% ee, Scheme 32). In addition, the AHF of α-methylstyrene was successfully performed at a Gram scale using only 0.05 mol% of the Rh-precursor and 0.15 mol% ligand for 90 h at 80 °C, to provide the desired chiral product in 92% yield and with 87% ee *(17)*.

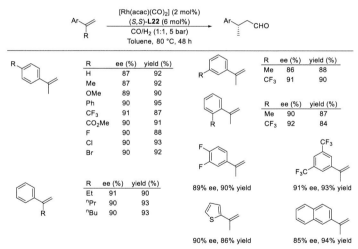

Scheme 32 Rh-catalyzed asymmetric hydroformylation of α-methylstyrene with (S,S)-DTB-YanPhos (**L22**).

More recently, the use of (S,S)-DTB-YanPhos ligand (**L22**, Scheme 25) was also reported in the efficient Rh-catalyzed asymmetric cyanide-free hydrocyanation of some 1-aryl-1-alkyl alkenes *(38)*. The transformation consists in the asymmetric hydroformylation/condensation/aza-Cope elimination sequence. The reaction of α-methyl styrene yielded the corresponding chiral nitrile with 86% ee and 88% yield. The high enantioselectivity was maintained for substrates with a longer alkyl chain and with electron withdrawing and donating groups in *para*-position of the benzyl ring (ee's up to 90%) (Scheme 33).

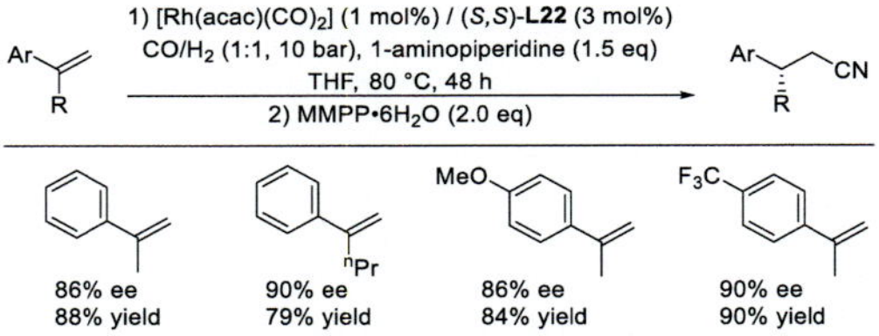

Scheme 33 Rh-catalyzed asymmetric cyanide-free hydrocyanation of 1-aryl-1-alkyl alkenes via tandem hydroformylation/condensation/aza-Cope elimination.

Dangat and Sunoj have recently performed a comprehensive DFT study on the asymmetric hydroformylation of α-methylstyrene catalyzed by the Rh/(S,S)-DTB-YanPhos (**L22**) system, with the aim to unravel the origin of the attained high regio- and enantioselectivities *(39)*. These calculations pointed out that non-covalent interactions (NCI) between the N-benzyl group and the other aromatic moieties in the ligand backbone are crucial for reaching good enantioselectivities (Fig. 2). Those interactions, in particular CH⋯π and π⋯π, seem to stabilize the most preferred conformers, giving an

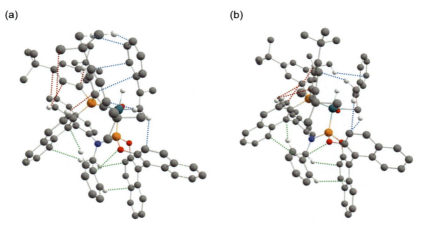

Fig. 2 Transition state geometries for the migratory insertion with coordination of the substrate through the *si*- (A) and *re*-faces (B). The figure shows the found NCIs. Ligand-ligand interactions are shown in red, substrate-ligand in blue and Bn group-ligand in green. Hydrogen atoms, wherever possible, are omitted to ensure better clarity.

appropriate shape to the chiral cavity to accommodate the incoming substrate. Moreover, NCIs between the chiral ligand and the substrate are also important since they favor a *si*-face binding, which leads to the (*S*)-linear product obtained experimentally. Fig. 2 shows the transition state geometries for the migratory insertion with coordination of the substrate thorough the *si*- and *re*-faces (Fig. 2A and B, respectively). The number of NCIs between the catalyst and the substrate (shown in blue dotted lines) is higher in the *si*-transition state (Fig. 2A), resulting in a higher stabilization, which is reflected by its lower relative free energy compared to that of the re-transition state.

4.1.3 1,1'-Diaryl olefins

So far, the AHF of 1,1'-diaryl olefins haven't receive much interest, which is reflected by the fact that only two papers describe its use as substrates *(25, 40)*. One study was already discussed above (see Section 3.2, Scheme 19), considering the *ortho*-pyridyl group of the substrate as an anchoring group to the metal center *(25)*. Another example was reported in 2002 by Botteghi and coworkers, who studied the asymmetric synthesis of the anti–muscarinic Tolterodine compound using as a key step the AHF of 1-[(2-hydroxy-5-methyl)]-1-phenylethene. Unfortunately, only 8% ee was obtained using (*R,S*)-Binaphos (**L3**) as the chiral ligand (Scheme 34). Moreover, high pressure and very long reaction time were needed. Moreover, high amounts of the hydrogenated product were obtained (27%) *(40)*.

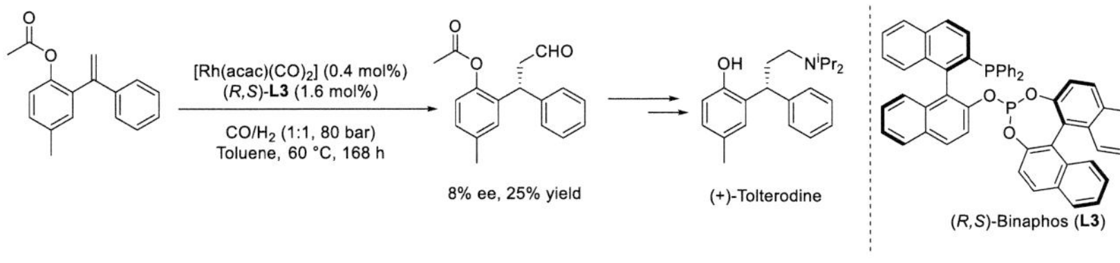

Scheme 34 Rh-catalyzed asymmetric hydroformylation of a diarylethene derivative with (*R,S*)-Binaphos (**L3**).

5. Rationalization of the catalyst efficiency

Achieving high chemo-, regio- and enantioselectivities on the AHF of 1,1'-disubstituted alkenes is not a trivial task. To deal with all those constraints, the selection of the catalyst is a must. Rhodium, despite its high cost, is more efficient in hydroformylation than platinum. Concerning ligands, it was observed that small structural changes from one type of 1,1′-disubstituted olefin to another require significant changes in ligand structure and reaction conditions. In addition, earlier ligands that were efficient for mono- and 1,2-disubstituted olefins such as Binaphos and Kelliphite did not work well for 1,1′-disubstituted olefins. Therefore, an approach based on fine-tuning modular ligands can be advantageous to improve the outcome in the hydroformylation of 1,1′-disubstituted olefins. Organophosphorus compounds maintain a privileged position in the AHF of 1,1′-disubstituted olefins with bidentate ligands providing the best selectivities. Only a few monodentate ligands were developed for this process, although with little success. Table 1 shows the most outstanding ligands for the AHF of 1,1′-disubstituted olefins, providing access to the desired chiral aldehydes with high yields and ee's for a set of substrates.

Rh catalytic systems containing P-stereogenic diphosphines, namely BenzP* (**L7**) and QuinoxP* (**L8**) were the first in providing high catalytic performance. Both can hydroformylate α-substituted acrylates with high yields and enantioselectivities, and in the case of QuinoxP*, a tandem reaction involving the AHF of acrylates and the subsequent amine condensation and Rh-catalyzed hydrogenation steps (the direct HAM) was also reported. Furthermore, QuinoxP* was also used in the i-selective AHF of 3,3,3-trifluoroprop-1-en-2-yl acetate, allowing the synthesis of an aldehyde with a quaternary stereocenter. Later, the diphospholane Ph-BPE (**L5**) was used in the AHF of allyl phthalimides and provided also high i-selectivities with electron deficient 3,3,3-trifluoroprop-1-en-2-yl acetate and α-methylene-y-butyrolactone. An important breakthrough came with the use of the phosphine-phosphoramidite YanPhos ligands. Rh complexes bearing this ligand family provided for the first time the AHF of different type of substrates with high yields and enantioselectivities. Hence, using either **L12** or **L22** (which differ in the substituents on the phosphine group, Table 1), the efficient asymmetric hydroformylation of α-substituted acrylates, allylic alcohols and amines, and even challenging unfunctionalized substrates as α-alkylstyrene derivatives and dialkyl-1,1′-alkenes, was demonstrated although the last group of substrates requires the introduction of a disulfonyl auxiliary. Moreover, they are also efficient in tandem reactions involving AHF. Finally, a last group of sugar-based phosphite-phosphoramidite ligands (**L13**) has also recently shown their potential, unlocking the AHF α-acrylamides. Furthermore, Rh/**L13** catalytic system also promoted the intermolecular asymmetric hydroaminomethylation (HAM) of a wide range of α-acrylamides.

Table 1 Selection of the most successful ligands developed for the AHF of 1,1'-disubstituted alkenes until now.

Ligand					
(R,R)-BenzP* (**L7**)	(R,R)-QuinoxP* (**L8**)	(S,S)-Ph-BPE (**L5**)	(S,S)-DTBM-YanPhos (**L12**) Ar = 3,5-tBu-4-MeO-C$_6$H$_3$	(S,S)-DTB-YanPhos (**L22**) Ar = 3,5-tBu-C$_6$H$_3$	**L13**
Ligand type	P stereogenic diphosphine		Phospholane	Phosphine phosphoramidite		Phosphite phosphoramidite
Substrate	8 examples up to 94% ee	(in HAM) 25 examples up to 86% ee 91% ee	9 examples up to 95% ee 95% ee 92% ee	19 examples up to 96% ee X= CH$_2$; 75% ee X= O; 86% ee) 26 examples X= O; up to 93% ee) X= NAc; up to 86% ee)	20 examples up to >99% ee 20 examples; up to 92% ee	9 examples; up to 99% ee also in HAM (up to 86% e)

All the ligands collected on Table 1 have the common feature of having a bulky environment around the phosphorus center. Indeed, the poor results achieved with 1,1'-disubstituted olefins can be mainly attributed to the difficulty in differentiating the two pro-chiral faces, and the bulkiness of the coordinative groups probably help in this task. On the other hand, looking at the two newest groups of successful ligands, both contain an N-substituted phosphoramidite group (Yanphos ligands **L12/L22** and sugar-based ligands **L13**). Another common characteristic found in the P-phoshoramidite ligands is that both contain a chiral binaphthyl moiety on the phosphoramidite group. Moreover, in both ligand families, the optimal configuration of this biaryl moiety is the (S). The substituent on the phosphoramidite moiety plays also an important role on the enantioselectivity. In the case of sugar-based ligands **L13** it was found that the best enantioselectivities were achieved with bulky groups, such as cyclohexyl or (S)-CHMePh. In contrast, Yanphos ligands work better with a less bulky benzyl group.

6. Conclusions and outlook

Due to the difficulty in differentiating the prochiral faces of the alkene, the AHF of 1,1'-disubstituted alkenes is much less developed than the reaction with monosubstituted or 1,2-disubstituted substrates. In this respect, it was not until 2011 that the first catalyst that provided useful enantioselectivities was developed; this was the Rh/BenzP* catalyst for the AHF of α-substituted acrylates (>80% ee). Since then, new advances have been made and nowadays some Rh-catalysts can hydroformylate 1,1'-disubstitued alkenes with high enantioselectivities, although the substrate scope is still limited. The most important advances were made for α-substituted acrylates and to a lesser extent for α-acrylamides, allylic alcohols and unfunctionalized substrates as α-alkylstyrene derivatives and dialkyl-1,1'-alkenes. Interests are also shifting to obtain exclusively the less favorable branched aldehyde giving access to the highly appealing α-tetrasubstituted aldehydes. It was shown that the formation of these products is favored for certain substrates holding electron withdrawing substituents using sterically hindered ligands, but high enantioselectivities has only been described for a few substrates and not always with satisfactory catalytic performance.

A survey of the literature also shows that only a few chiral ligands provide effective rates and selectivities. The first successful ligands were P-stereogenic diphosphines and chiral diphospholanes until the phosphine-phosphoramidite YanPhos ligands appeared and allowed for the first time the AHF of different substrate types. Lately, sugar-based phosphite-phosphoramidite ligands also demonstrated their potential, unlocking the AHF of α-acrylamides.

To ensure the industrial applicability of a chemical process it becomes crucial to achieve catalytic systems that operate under mild reaction conditions. Gratifyingly, despite the very high CO/H_2 pressures required in the first attempts (typically 80–190 bar), nowadays it is possible to operate at a pressure as low as 5–20 bar. However, the reaction temperature required is still high, only few systems work at 60 °C and in most of the cases temperatures of 80 °C or higher are needed. This is a drawback not only for the sustainability of the reaction, but also because it complicates even more to reach high levels of asymmetric induction.

Acknowledgments

We gratefully acknowledge financial support from the Spanish Ministry of Economy and Competitiveness (PID2019-104904GB-I00 and PID2019-104427RB-I00), European Regional Development Fund (AEI/FEDER, UE), the Catalan Government (2017SGR1472), the ICREA Foundation (ICREA Academia award to M.D) and the European Union's Horizon 2020 research and innovation programme (Marie Skłodowska-Curie grant No 860322 to J.L).

References

1. (a) Franke, R.; Selent, D.; Börner, A. *Chem. Rev.* **2012**, *112*, 5675–5732; (b) Whiteker, G. T.; Cobley, C. J. Applications of rhodium-catalyzed hydroformylation in the pharmaceutical, agrochemical, and fragrance industries. In *Organometallics as Catalysts in the Fine Chemical Industry*; Beller, M., Blaser, H. U., Eds.; Topics in Organometallic Chemistry, Vol. 42; Springer: Berlin, Heidelberg, 2012.
2. *Oxo-Alcohol Market: Global Industry Trends, Share, Size, Growth, Opportunity and Forecast 2021–2026*; 2021. https://www.imarcgroup.com/oxo-alcohol-technical-material-market-report(Accessed 16 July 2021).
3. (a) Agbossou, F.; Carpentier, J. F.; Mortreaux, A. *Chem. Rev.* **1995**, *95*, 2485–2506; (b) Deng, Y.; Wang, H.; Sun, Y.; Wang, X. *ACS Catal.* **2015**, *5*, 6828–6837; (c) Cunillera, A.; Godard, C.; Ruiz, A. *Top Organomet. Chem.* **2018**, *61*, 99–144; (d) Brezny, A. C.; Landis, C. R. *Acc. Chem. Res.* **2018**, *51*, 2344–2354; (e) Chakrabortty, S.; Almasalma, A. A.; de Vries, J. G. *Catal. Sci. Technol* **2021**, *11*, 5388–5411.
4. (a) Shultz, C. S.; Krska, S. W. *Acc. Chem. Res.* **2007**, *40*, 1320–1326; (b) Ager, D. J.; de Vries, A. H. M.; de Vries, J. G. *Chem. Soc. Rev.* **2012**, *41*, 3340–3380; (c) Seo, C. S. G.; Morris, R. H. *Organometallics* **2019**, *38*, 47–65.
5. (a) Gonzalez, A. Z.; Román, J. G.; Gonzalez, E.; Martinez, J.; Medina, J. R.; Matos, K.; Soderquist, J. A. *J. Am. Chem. Soc.* **2008**, *130*, 9218–9219; (b) Corberán, R.; Mszar, N. W.; Hoveyda, A. H. *Angew. Chem., Int. Ed.* **2011**, *50*, 7079–7082.
6. Keuelemans, A. I. M.; Kwantes, A.; van Bavel, T. *Recl. Trav. Chim. Pays-Bas* **1948**, *67*, 298–308.
7. Clarke, M. L.; Roff, G. J. *Chem. Eur. J.* **2006**, *12*, 7978–7986.
8. Ning, Y.; Chen, F.-E. *Green Synth. Catal* **2021**, *2*, 247–266.
9. Kollár, L.; Consiglio, G.; Pino, P. *J. Organomet. Chem.* **1987**, *330*, 305–314.

10. Parrinello, G.; Stille, J. K. *J. Am. Chem. Soc.* **1987**, *109*, 7122–7127.
11. Kollár, L.; Consiglio, G.; Pino, P. *Chimia* **1986**, *40*, 428–429.
12. Wang, X.; Buchwald, S. L. *J. Am. Chem. Soc.* **2011**, *133*, 19080–19083.
13. Cunillera, A.; de los Bernardos, M. D.; Urrutigoïty, M.; Claver, C.; Ruiz, A.; Godard, C. *Catal. Sci. Technol.* **2020**, *10*, 630–634.
14. Jia, X.; Ren, X.; Wang, Z.; Xia, C.; Ding, K. *Chin. J. Org. Chem.* **2019**, *39*, 207–214.
15. Li, S.; Li, Z.; You, C.; Li, X.; Yang, J.; Lv, H.; Zhang, X. *Org. Lett.* **2020**, *22*, 1108–1112.
16. You, C.; Li, S.; Li, X.; Lv, H.; Zhang, X. *ACS Catal.* **2019**, *9*, 8529–8533.
17. You, C.; Li, S.; Li, X.; Lan, J.; Yang, Y.; Chung, L.; Lv, H.; Zhang, X. *J. Am. Chem. Soc.* **2018**, *140*, 4977–4981.
18. Noonan, G. M.; Newton, D.; Cobley, C. J.; Suásrez, A.; Pizzano, A.; Clarke, M. L. *Adv. Synth. Catal.* **2010**, *352*, 1047–1054.
19. Miró, R.; Cunillera, A.; Margalef, J.; Lutz, D.; Börner, A.; Pamiès, O.; Dieguéz, M.; Godard, C. *Org. Lett.* **2020**, *22*, 9036–9040.
20. McDonald, R. I.; Wong, G. W.; Neupane, R. P.; Stahl, S. S.; Landis, C. R. *J. Am. Chem. Soc.* **2010**, *132*, 14027–14029.
21. Domke, L.; PhD dissertation, Universität Rostock, Germany, 2015.
22. Nozaki, K.; Li, W.-G.; Horiuchi, T.; Takaya, H. *Tetrahedron Lett.* **1997**, *38*, 4611–4614.
23. Bäuerlein, P. S.; Gonzalez, I. A.; Weemers, J. J. M.; Lutz, M.; Spek, A. L.; Vogt, D.; Müller, C. *Chem. Commun.* **2009**, 4944–4946.
24. Zheng, X.; Cao, B.; Liu, T. L.; Zhang, X. *Adv. Synth. Catal.* **2013**, *355*, 679–684.
25. Basoli, C.; Botteghi, C.; Cabras, M. A.; Chelucci, G.; Marchetti, M. *J. Organomet. Chem.* **1995**, *488*, C20–C22.
26. Lee, C. W.; Alper, H. *J. Org. Chem.* **1995**, *60*, 499–503.
27. Wang, X.; Buchwald, S. L. *J. Org. Chem.* **2013**, *78*, 3429–3433.
28. Tan, R.; Zheng, X.; Qu, B.; Sader, C. A.; Fandrick, K. R.; Senanayake, C. H.; Zhang, X. *Org. Lett.* **2016**, *18*, 3346–3349.
29. Eshon, J.; Foarta, F.; Landis, C. R.; Schomaker, J. M. *J. Org. Chem.* **2018**, *83*, 10207–10220.
30. Gladiali, S.; Faedda, G.; Botteghi, C.; M, M. J. Organomet. Chem. 1983, 244, 289–302.
31. Consiglio, G.; Nefkens, S. C. A.; Borer, A. *Organometallics* **1991**, *10*, 2046–2051.
32. Zhang, D.; You, C.; Li, X.; Wen, J.; Zhang, X. *Org. Lett.* **2020**, *22*, 4523–4526.
33. Consiglio, G.; Morandini, F.; Scalone, M.; Pino, P. *J. Organomet. Chem.* **1985**, *279*, 193–202.
34. Consiglio, G.; Roncetti, L. *Chirality* **1991**, *3*, 341–344.
35. Kollár, L.; Bakos, J.; Tóth, I.; Heil, B. *J. Organomet. Chem.* **1988**, *350*, 277–284.
36. Ojima, I.; Takai, M.; Takahashi, T. WO Patent 078766, 2004.
37. Carrilho, R. M. B.; Neves, A. C. B.; Lourenço, M. A. O.; Abreu, A. R.; Rosado, M. T. S.; Abreu, P. E.; Eusébio, M. E. S.; Kollár, L.; Bayón, J. C.; Pereira, M. M. *J. Organomet. Chem.* **2012**, *698*, 28–34.
38. Li, X.; You, C.; Yang, J.; Li, S.; Zhang, D.; Lv, H.; Zang, X. *Angew. Chem. Int. Ed.* **2019**, *58*, 10928–10931.
39. Dangat, Y.; Popli, S.; Sunoj, R. B. *J. Am. Chem. Soc.* **2020**, *142*, 17079–17092.
40. Botteghi, C.; Corrias, T.; Marchetti, M.; Paganelli, S.; Piccolo, O. *Org. Process Res. Dev.* **2002**, *6*, 379–383.

About the authors

Dr. Jèssica Margalef received her PhD in 2016 at the University Rovira i Virgili (Tarragona) under the supervision of Prof. Montserrat Diéguez and Dr. Oscar Pàmies. During her PhD she did a 3 months exchange in the group of Prof. Hans Adolfsson (Stockholm University) and a short stage in Prof. Per-Ola Norrby (Gothenburg University). In January 2017, she joined Prof. Joseph Samec's group at Stockholm University as a postdoctoral researcher. After two years, she came back to Tarragona as a Martí Franquès fellow. Her research interests include homogeneous catalysis and DFT-guided development of new catalysts.

Joris Langlois completed his BSc in chemistry from University of Caen Normandy, France. He obtained his MS degree in Molecular and Supramolecular chemistry from the university of Strasbourg and his MS degree in chemical engineering from the European school of Chemistry, Polymers and Materials sciences (ECPM, Strasbourg). He did internships in medicinal chemistry with Servier pharmaceuticals (Paris and Budapest) and in polymer chemistry with L'Oréal (Paris). He joined the CCIMC Project (European Joint Doctorate) as an early-stage researcher in 2020. He is supervised by Dr. Cyril Godard in the University Rovira i Virgili and Prof. Martine Urrutigoïty in the Institut National Polytechnique of Toulouse. His research focuses in asymmetric hydroformylation and asymmetric hydroaminomethylation of alkenes.

Guillem Garcia obtained his Bachelor's Degree in Chemistry in 2021 at the University Rovira i Virgili (Tarragona). During his last year he obtained a collaboration grant that allowed him to work in the group of Prof. Diéguez. He is currently studying a master's degree in synthesis, catalysis and molecular design at the same university. His research interests include asymmetric homogeneous catalysis.

Dr. Cyril Godard received his PhD degree in in Inorganic Chemistry from the University of York, UK, in 2004 under the supervision of Profs. R. Perutz and S. Duckett for his work on bond activation reactions at Rh cyclopentadienyl complexes. After a postdoctoral position in York working on Para-Hydrogen Induced Polarisation techniques, he was awarded the Juan de la Cierva Fellowship in 2006 and the Ramon y Cajal fellowship in 2009 at the Universitat Rovira i Virgili in Tarragona, Spain, where he is currently associate professor in Inorganic Chemistry. His research interests lie in the areas of homogeneous catalysis and application of transition metal nanoparticles in catalysis. He is currently the co-author of 82 scientific articles in international journals and 9 book chapters (h-index = 24).

Prof. Monterrat Diéguez got her PhD in 1997 at the Rovira i Virgili University (URV). She was post-doc at Yale University with Prof. R.H. Crabtree. Since 2011 she is full professor in Inorganic Chemistry (URV). She is the chair of InnCat research group at URV, succeeding the former chair, Prof. Claver. She is author of 160 articles and 18 books/book chapters with an H index of 42. She got the Distinction from the Generalitat de Catalunya in 2004 and in 2008 from the URV. She got the ICREA Academia Prize in 2009–14 and 2015–20. Her research interests are catalysis, combinatorial synthesis, artificial metalloenzymes, and catalytic conversions of renewable feedstocks.

Printed in the United States
by Baker & Taylor Publisher Services

A Minimally Good Life